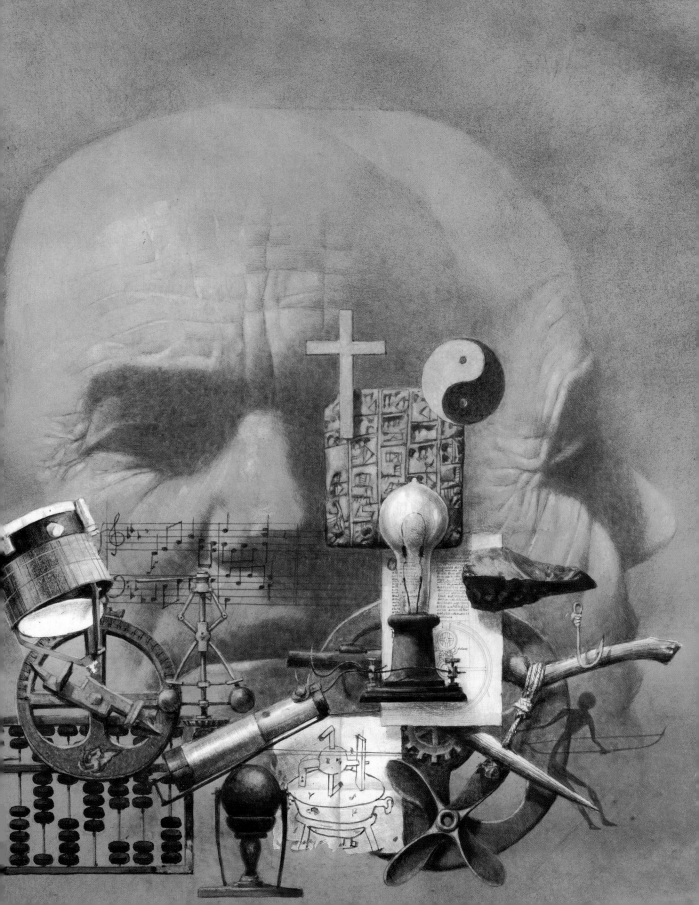

IDEAS THAT CHANGED THE WORLD

ART AND TECHNOLOGY
THROUGH THE AGES

ILLUSTRATED BY ROBERT INGPEN
TEXT BY
PHILIP WILKINSON & JACQUELINE DINEEN

CHELSEA HOUSE PUBLISHERS
New York • Philadelphia

First published in the United States in 1995 by
Chelsea House Publisher

First Printing
1 3 5 7 9 8 6 4 2

Simplified text and captions by **Jacqueline Dineen**
based on the *Encyclopedia of Ideas that Changed the
World* by Robert Ingpen & Philip Wilkinson

Editor	Diana Briscoe
Project Editor	Claire Watts
Designer	Design 23
Art Director	John Strange
Design Assistants	Karen Fergusom
	Victoria Furbisher
DTP Manager	Keith Bambury
Editorial Director	Pippa Rubinstein

ISBN 0–7910–2769–4

Printed in Italy

CONTENTS

◆

Introduction

Reading and writing seem the most natural ways to comunicate in the world. Everywhere we look – on road signs, posters, computer screens – there are words to be read. And using satellites and fax machines, radio and television, words and pictures can be transmitted all the way around the world.

But for early people, it was a different story. Our earliest ancestors could not write at all, although some of their most ancient remains, paintings in caves and on rocks, are of images that may have held deep meanings for them. Then, about 5,500 years ago, the Sumerian people of the Middle East devised the first writing system. Beginning with simple pictures of the objects they wanted to record, they developed a system of signs. Other ancient peoples, such as the Egyptians, Phoenicians, and Chinese, created their own writing systems. Soon everything from business deals to poetry could be written down on materials as different as clay tablets and papyrus scrolls.

Even so, all books had to be written by hand. Every word in every volume had to be copied out laboriously by a scribe. The thing that changed all this, and that brought about the first great revolution in communications, was the invention of printing with movable type around 1450. Suddenly, books could be produced cheaply and quickly. Printing presses spread through Europe, and beyond, and their products were sold far and wide.

It was not until the middle of the nineteenth century, with inventions like the telegraph and telephone, that there was an advance of similar importance. Then, in the late 1890s, came the still more amazing development that allowed messages to be sent through the air – radio.

The printing press had allowed people to read the words of a writer who lived hundreds, perhaps thousands, of miles away. Radio did more, allowing you to listen to a distant speaker. Television would later provide a similar way of carrying moving images.

It is difficult to imagine today how amazing these inventions seemed to the people of the time. But one invention has been even more important – the computer. Computers began as a way of doing calculations quickly. They are now used for everything from playing games to servicing cars. They are also used in all kinds of ways in communications, to the point where the "electronic book" is likely to cause changes as vast as those brought about by the printing press.

PHILIP WILKINSON

THE FIRST ART

*Once early people had learned
the skills they needed for survival, they
began to look around them and record the
things they saw as pictures and models.*

Human beings have evolved over four million years, but our own direct ancestors, *Homo sapiens sapiens*, appeared only 35,000 years ago. The period from the emergence of *Homo sapiens sapiens* to the time when people began to settle in groups and farm the land is known as the Old Stone Age. The people of the Old Stone Age were hunter-gatherers who had spread from Africa, which is thought to be the birthplace of the first humans. They had moved all over the world, to Europe, Asia, Australia, and the Americas. These people could make a variety of tools and had learned to light a fire and to build simple shelters. Once they had learned how to keep themselves alive, they began to express themselves by drawing the things they saw around them. This was the first art.

Some of the earliest art appears on the walls of caves where people sheltered at night. Paintings, engravings, and carvings show the animals that people hunted. The most common are bison, deer, oxen, and

△ *Owl wall-painting from northern Australia.*

horses, but there are also paintings of lions, bears, mammoths, birds, and fish in some caves.

Why did people start to paint the walls of caves in this way? Was it simply decoration, meant to represent the world around them in pictures, or was it part of some ceremony or ritual? And how did these early artists realize that they could draw, when no one had done so before?

Some pictures were engraved on top of each other in a confusing crisscross of lines. Sometimes it is still possible to see the outline of an animal.

SEEING SHAPES

Creating art is part of the human instinct. Small children start to draw simple images from an early age and these become more realistic as they gain greater understanding of the world around them. The ability to draw seems to have evolved

▽ *Animal hunt cave painting.*

as people became more advanced. Perhaps it began as they looked around the caves they sheltered in. The insides of the caves were dark and mysterious. People only had the light of primitive lamps to see by. At first, they may have simply begun to see shapes suggested by bumps and patterns in the rock, in the same way that we can sometimes see shapes in clouds when we look up at the sky. Maybe an outline in the rock suggested a mammoth or a deer, and someone decided to improve on nature by scratching in more detail. From there, they could move on to drawing the complete animal and to shaping it in other ways, such as modeling with clay, or carving bone or antler.

Examples of early art have been found in Africa, Asia, the Americas, and Australia, but the finest evidence comes from Europe, particularly from caves in France and Spain. There are examples of "portable" art in the caves, such as carvings in stone, bone, and wood, engravings on pieces of stone and bone, and models molded in clay. One of the most famous clay models is the pair of bison in a cave at Le Tuc d'Audoubert in France. Each animal is almost 40 inches (1 meter) long, about one-sixth of its real size.

DRAWING ON THE WALLS

On the walls themselves, paintings or engravings are carved into the rock wall. The engravings are sometimes difficult to make out because they have not been carved very deeply. The artists would

△ *The artists of the Ice Age used sticks to mix their paint and often put it on the walls with their bare hands. They produced some of the finest cave paintings. Often a whole range of different subjects was represented in a cave.*

have needed a range of tools to carry out the carving. To carve fine lines, they used a tool with a sharp point and edge called a "burin." They used a sharp pointed tool called an "awl" for piercing hides. Larger, stronger tools such as flint blades and hammers could be used for chipping away bigger pieces of rock.

Some of the most beautiful works are the paintings. These early artists used natural materials to make their paints. Ocher is a kind of earth made up of clay and minerals. It provided red, yellow, and brown pigments, and charcoal provided black. No evidence of blue or green paint has been found. The pigments were

mixed with water to make paint which was put onto the wall with the artist's hands or with a twig or stick or a brush made from animal hair. Some pictures were a mixture of carving and painting, making use of natural cracks and bumps in the wall. Artists also used a simple stenciling technique. Someone placed his hand against the wall and the artist sprayed paint onto it. When the hand was taken away, its shape was left on the wall.

The paint was probably sprayed on with a simple blowpipe made of bone.

MAGIC RITUALS

But what were these pictures all about? People are hardly ever shown in them. When pictures do include people, they are often wearing masks, animal skins, and antlers on their heads. These could have been disguises for hunting, but there is nothing to show that the cave paintings

In ancient times, sometimes giant figures of animals or men were carved into the ground, so that they could be seen from a long way off.

▽ *Prehistoric art has been found all over the world, but the best examples come from Europe. From left to right:*

1 *Charging bison, wall painting, Altamira, Spain, c. 20,000 B.C.*
2 *Venus of Willendorf, limestone carving, Germany, c. 30,000 B.C.*
3 *Female figure, ivory carving, central Europe, c. 20,000 B.C.*
4 *Head of girl, ivory carving, Brassempouy, France, c. 20,000 B.C.*
5 *Female figure, carving, Germany, c. 20,000 B.C.*
6 *Staff or spear thrower, antler carving, southern France, c. 20,000 B.C.*
7 *Statue, stone carving, St. Sernin, France, c. 3000 B.C.*

were about hunting itself, so it is more likely that they were part of some religious ritual. The paintings are often in parts of caves that are very hard to reach, hidden away and dark. The artists would have needed ladders to get to them. Simple ladders were probably made from a small tree trunk with the branches trimmed down. The artists would also have needed lights to work by, but they chose to paint here rather than on walls nearer the mouth of the cave.

SUCCESSFUL HUNTING

The fact that the paintings were hidden away like this suggests that the caves may have been used for secret rituals. Perhaps the artists simply felt that drawing pictures of animals would make the hunt more successful, or perhaps they were drawn as part of a ceremony that took place before the hunt.

Some of the animals have been painted with arrows stuck in them, others look as though they have been attacked with stones. Perhaps the people believed that this would help them to kill real animals. Or perhaps the artists drew animals to make sure that their numbers increased so that there would always be plenty to hunt.

We cannot really know the reasons behind the paintings. Maybe, like art today, they were carried out for different reasons. Some may have been the work of artists trying to paint what they saw around them. Others may have had religious significance. But whatever their reasons for their paintings and carvings, artists went on adding to the cave art for about 15,000 years.

HUMAN FIGURES

Most early art portrays animals but some statues of women have been found as well. The most famous of these are the "Venus" figures. They are all fat, with large breasts, a rounded stomach, and massive thighs. Perhaps prehistoric man's idea of female beauty was a fat, curvaceous figure, or maybe these figures are meant to represent goddesses.

One of the best-known is the plump, well-rounded figure of the Venus of Willendorf, which was carved in Germany about 30,000 years ago. The Venus of Laussel was carved in the Dordogne area of France at about the same time.

Other female statues are slim or seem to be of older women. Archaeologists are not sure what these figures represent. Some female figures have been found in special pits in what were once the floors of huts. They may have been models of the inhabitants' ancestors, as some early people worshiped their ancestors.

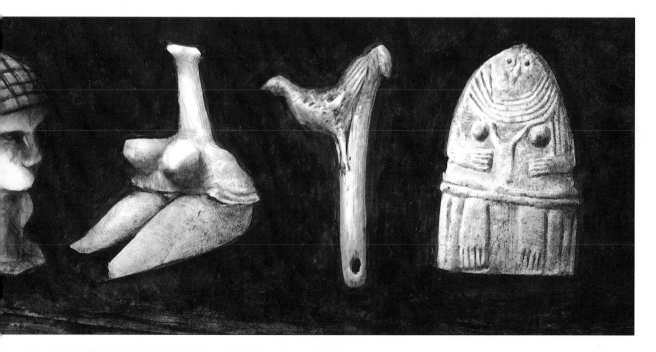

POTTERY

Pottery was one of the first skilled crafts that early people discovered. As the demand for pottery containers and dishes grew, some of the earliest machines and manufacturing techniques were developed.

W hen people began to settle in groups and farm the land, they needed containers for storing grain and other crops. People had already noticed that clay can be shaped when it is wet and sticky, and that it bakes hard when it is heated. They used this discovery to build houses with mud bricks baked in the sun, and they soon also began to make simple clay pots.

Pottery-making is one of the oldest crafts. The earliest pottery found is from the Far East and dates back to 10,000 B.C.

In the West, it seems to have appeared in Africa and the Near East first and spread into Europe by about 4000 B.C.

The first pots were made by simply shaping a lump of clay by hand, or by coiling. The base of the pot was made from a flat disk of clay or a clay strip rolled round into a cartwheel shape.

△ *Early Chinese vessel.*

▷ *The invention of the wheel meant that potters could make more elaborate pots far more quickly and easily. Simple potter's wheels like this one are still used in many parts of the world today.*

△ *Early pottery was made either by shaping clay with the thumbs, by molding it, or by building up coils of clay.*

More clay was rolled into a long strip and coiled around on the edge of the base to build up the sides. When the pot was finished, the surface was smoothed down to give it a flat surface. It could then be decorated by scratching the soft clay or by painting it when it was dry.

Flat dishes could be shaped by pressing clay into a mold such as a basket or an existing dish. Another simple method was to mold clay around a suitably shaped stone and hammer it into shape.

When the pots had been shaped, they had to be baked or "fired," so that the water in the clay evaporated and it became hard. The most obvious method was to dry the pots in the sun. This system worked in hot parts of the world, but the cooking fire made a more reliable source of heat in cooler climates.

Once people had learned to shape and fire clay in this way, each family could produce its own pottery. Some potters would have been more skillful than others, and perhaps began to design more imaginative and decorative pottery. It was not long before specialized potters began to concentrate on this work.

EARLY KILNS

Baking clay in the sun or in the embers of the fire is satisfactory for making pots to hold dry produce such as grain, but it has disadvantages. The heat of the sun or an open fire is not great enough to make the clay completely watertight. It is still porous like a sponge, so water soaks into it. The answer is to fire it in a closed oven, or "kiln," which can reach a higher temperature to make the surface of the clay melt before it hardens. The molten clay forms a watertight skin.

Kilns dating back to about 4500 B.C. have been found in China. They were used in Mesopotamia by about 4000 B.C. and in Egypt about 1,000 years later. These early kilns consisted of two chambers, one above the other. The lower chamber held the fire and the upper chamber held the pots. The two chambers were connected by a vertical "flue" or chimney. This intensified the heat by forcing it up into the pottery chamber instead of allowing it to disperse in all directions. The pottery had to be stacked carefully so that the heat could circulate around the chamber and reach each pot. The shape of the kiln was important. The firing chamber was dome-shaped with the flue in the center of the roof. This shape helped to direct all the heat up the flue as efficiently as possible.

THE POTTER'S WHEEL

Early potters realized that their work would be made easier if the pot was turned as they made it. The first step was to put the pot on a disk fixed on to a pivot, which was a piece of wood driven into the ground. The potter could turn the disk with one hand and shape the pot with the other. This simple potter's wheel was low and uncomfortable for the potter to work on, so the disk was raised up on a taller shaft and set into a stone or wooden base. This device, known as the "tournette," first appeared in Mesopotamia in about 3500 B.C.

It took another 2,000 years for the next development, the true potter's wheel. A second disk was fitted to the bottom of the shaft, enabling the potter to turn the wheel with his feet. The potter could now spin the wheel much faster and also had both hands free to shape the pot.

Once potters had discovered these methods, they could develop their art and create beautiful objects, some of which can still be seen today.

PORCELAIN

Porcelain is a fine, white, delicate type of china made from a special clay called "kaolin." Kaolin was first used to make porcelain in China which is why we know it as "china." The Chinese had invented a kiln which could be heated to very high temperatures. They found that kaolin became hard and glass-like when it was fired at these temperatures. This process is called "vitrification." Adding a mineral called "petuntse" helped by lowering the temperature needed to start vitrification. It also gave the porcelain its particular shine.

The Chinese first began to make porcelain in about A.D. 900 and for centuries they kept its secret to themselves. Their wonderful vases, with their delicate texture and ornate decorations, became one of the wonders of the Orient.

For centuries, European potters tried to make porcelain but they could not discover the secret. Then, in the eighteenth century, a German alchemist named Johann Böttger (1682–1719) began a series of secret experiments in an attempt to unravel the mystery. He discovered the method in 1708, and the first European porcelain factory was set up at Meissen. Some of the world's finest porcelain still comes from Meissen.

GLASS

It is difficult for us to imagine that hard, transparent glass is made from tiny grains of sand. Imagine how much more difficult it was to discover this fact in the first place, and develop the techniques for manufacturing it!

I t was quite simple for people to notice and begin to make use of the properties of clay, but it was not so easy to find out about glass, which is made from a mixture of substances.

The main ingredient of glass is "silica" or sand. If silica is heated to a very high temperature, it melts into drops of glass. When several of these drops fuse together, they form liquid glass which can be shaped. The glass hardens when it cools. Silica will only melt at a temperature of 2,700°F (1,500°C) or more, and it is difficult and expensive to maintain such intense heat. Glass made from pure silica is very brittle and breaks easily, so in glassmaking, soda ash and crushed limestone are added to the silica to lower the melting point and harden the glass.

CARVING AND SHAPING

By about 2600 B.C., glassmakers in Mesopotamia had learned to make glass. These early attempts were crude because

△ *Early glass vessels made in Mesopotamia were made by carving glass rather than by blowing it.*

▽ *To blow glass, air is blown along a hollow tube into a piece of molten glass attached to one end.*

early furnaces were not very efficient, and it was difficult for glassmakers to melt the ingredients properly. The Mesopotamians did not work their glass while it was soft. They carved dishes and bowls from solid lumps of hardened glass, in the same way as carving stone.

The Egyptians seem to have been the first to master the art of making beautiful glass objects, as we can see from vessels found in royal tombs. Egyptian glassmakers made vessels of patterned glass in different colors. They were shaped by building up layers of molten glass on a clay cast of the vessel. The clay

shape was attached to a metal rod which the glassmaker held as he added threads of different colored glass. He made patterns by pulling the threads into zigzags or wavy lines with a special tool shaped like a comb.

GLASSBLOWING

The technique of glassblowing for shaping items such as drinking glasses and bottles was invented in about 100 B.C., probably in Syria. A glassmaker must have found that if he blew down a hollow tube with a blob of molten glass on the end, the air made the glass inflate like a balloon. The early glassmakers learned to shape bottles and glasses by blowing the glass into a mold.

A ROMAN INDUSTRY

This method of shaping hollow vessels was soon adopted by the Romans, who had the largest glass industry of any civilization in the ancient world. The Romans produced exquisitely decorated vases in glowing colors as well as everyday items such as jars and bottles. Glassware and the technique of making it spread throughout the Roman Empire, and huge quantities of glass were produced for trading.

Glass is easily broken and it can also be melted down and used again so relatively few examples have survived from ancient times. In excavations at the Roman towns of Pompeii and Herculaneum, which were buried under ash and lava after the eruption of Vesuvius in A.D. 79, archaeologists have discovered glass

bottles, beakers, plates, jugs, vases, and delicate bottles for cosmetics, designed in a range of styles. From the second to the fourth centuries, glassmakers produced expensive styles such as cut-glass and vessels decorated with white figures in relief on a darker background, or with ribbing in delicate patterns.

Glass was made throughout the Roman Empire. Some pieces have a molded trademark which tells archaeologists where they were made. For example, the trademark CCA (Colonia Claudia Agrippinensis) tells us that the piece was made in Cologne, Germany.

WINDOW GLASS

Early glass was full of impurities, and glassmakers could only produce fairly small pieces. The methods of making sheet and plate glass for windows came much later. Early window glass was made in small pieces joined together with strips of lead, but it was a laborious and

Glass can be made in many colors and is easy to shape.

expensive process. Windows with glass in them were only seen in the homes of the nobility until after the Middle Ages.

The technique of glassmaking, spread by the Romans, continued to develop. By the sixteenth century, many large towns had glassworks and a variety of beautiful objects were being produced. Venice, Italy, had become the home of exquisite glass and Venetian glass is still admired today. Designs from Venice inspired glassmakers elsewhere to make the delicate and fragile ornaments that graced the homes of the rich. Glass also had more practical uses. For example, vessels were specially made for carrying out scientific experiments. By now, glassmaking processes were improving. Glass was more transparent and had fewer impurities.

Glassblowing by machine was introduced during the Industrial Revolution, which began in about 1750.

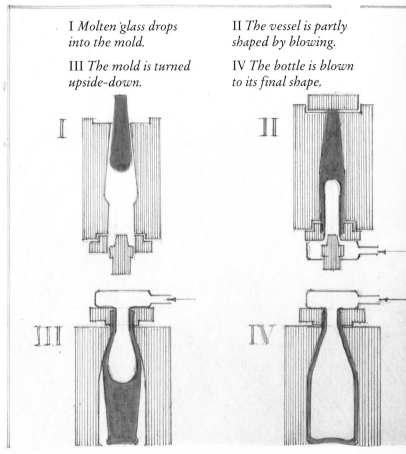

I *Molten glass drops into the mold.*

II *The vessel is partly shaped by blowing.*

III *The mold is turned upside-down.*

IV *The bottle is blown to its final shape.*

SHEETS OF GLASS

Early glassmakers could not make glass in large, flat sheets. An early method was to pull molten glass up the side of a tower and allow it to harden. The glass produced could be used for windows, but it was very distorted and hard to see through. The next development was "plate glass" which was made by pouring molten glass on to a table and rolling it flat. The glass had to be polished to remove the roller marks which was an expensive process. But plate glass was the only answer for making large sheets of window glass until 1959, when the "float-glass" method was invented. Molten glass is floated on a bath of molten tin, producing perfectly smooth, flat glass.

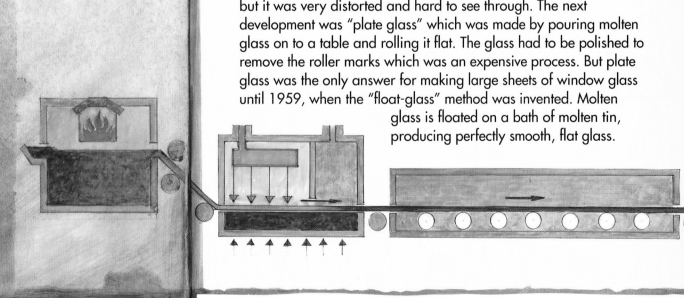

METAL

The discovery of metal opened up a whole range of possibilities to early man. Here was a hard material that could be melted, molded, and shaped into a whole variety of strong, sharp, new tools.

About 9,000 years ago, a group of craft-workers made a discovery that was to change the world. They realized that metal was a natural mineral that could be melted into a liquid by heat. For centuries, stone had been the only material people had for making tools, weapons, and other implements. Stone could only be shaped by carving, and this was clumsily done with simple tools.

Metal's great advantage over stone was that, when molten, it could be poured into molds and shaped in a variety of intricate ways. When it cooled, it became hard and strong, and blades could be sharpened until they were lethal.

The discovery of metal was probably an accident, like so many of the things early people learned about. People came across lumps of strange material, which

△ *Metals were not just used for tools but for jewelry and ornaments too.*

▷ *Metals were smelted in stone furnaces lined with clay. The molten metal ran out at the bottom of the furnace. It could be used to make tools such as saws with fine teeth and sharp swords which were unknown before.*

MINING FOR METALS

The first lumps of metal were found lying around in rivers or on the Earth's surface. These are known as "free metals." Then, people found "metallic ores," or rocks which contained metals. They began to realize that there could be more metallic ores buried beneath the ground.

The earliest known copper mines are in southeastern Europe. They are about 7,000 years old. The miners dug shafts about 65 feet (20 meters) deep and removed the ore by "fire-setting." They lit fires to melt the metal in the rock, then threw water on to it to make it cool suddenly. This split the rock, and the miners could then drive wooden wedges in to remove the ore. The broken lumps were smashed into pieces to remove as much rock as possible before smelting.

The same system was used in copper mines in Salzburg, Austria, in about 1500 B.C. As ore was removed, the miners dug tunnels into the rock face and supported them with wooden ceilings. These tunnels helped the air to circulate so that the fires would burn properly.

felt quite different from the types of stone they were used to. They may have been searching for new types of clay or stone to make pots with. Perhaps they tried to carve the new material that they had discovered with their bone and flint tools and found it impossible. Then someone had the idea of beating it with a hammer. It did not break like stone. Instead, the hammering made the substance change its shape.

HEAT AND HAMMERS

The first metals to be found were copper and gold. Gold can be hammered into shape easily, but copper is brittle and breaks if it is hammered too much. The next discovery was that copper could be shaped without breaking if it was first hammered, then heated and hammered again. This process is called "annealing."

Then the craft-workers found that if they put these metals into their pottery

PATTERN WELDING

Cast iron was more brittle than wrought iron, so it was not ideal for making weapons such as swords which could snap in a fight. Metalworkers used a technique called "pattern welding" to make many weapons and tools.
They made strips of cast iron and then hammered them together.
They could build up the shape they wanted, and the hammering strengthened the iron.

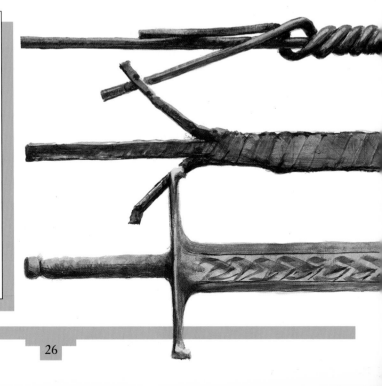

kilns, they melted into liquid. A potter may have decided to put a lump of metal into the kiln just to see what happened, or perhaps they tried to experiment with metals as coloring or decoration for their pottery. Kilns may have been dug in ground where there were lumps of rock which contained metal. The metal would have melted and dripped to the bottom of the kiln.

This discovery was important in another way. People began to realize that if they could find rock with metal in it, the two could be separated by heating the rock until the liquid metal flowed out. This process, known as "smelting," was discovered by metalworkers in the Middle East in about 7000 B.C.

THE BRONZE AGE

Copper was probably the first metal to be smelted. As the search for metals progressed, silver, lead, and tin were discovered. The metalworkers realized that these metals all had different properties which made them harder or softer, stronger or more brittle than each other. Copper was an attractive metal for making ornaments but it was soft and dented easily, so it was not much use for making tools or weapons. But could it be made stronger by adding another metal with different properties to the copper? As metalworkers began to experiment with mixing metals together, they discovered that adding tin to copper made a metal that was both attractive and strong. Bronze, as this new metal was called, was the first "alloy," or artificial metal made by mixing two or more natural metals together.

Now metalworkers could make bronze spears, swords, axes, and other weapons. These metal weapons had an advantage

▽ *Three iron strips are twisted together and hammered into place. The sword is gradually hammered into shape until it is strong and sharp.*

over stone. Flint spearheads and arrowheads had to be fixed to wooden shafts with twine, but metal could be shaped so that the head fitted into a socket in the shaft.

Copper was used over a large part of Europe and Asia by about 3000 B.C. In Asia, bronze was being used as well. By about 2000 B.C., most metalworkers had switched to bronze. Bronze weapons were in common use throughout most of Europe and western Asia by 1000 B.C., but the ingredients to make bronze were in short supply. The main sources of copper were Austria and the Balkans, and tin was mainly found in Britain, France, Spain, and northern Italy. There was very little of either metal in the Middle East where many of the large cities were, so metals had to be brought along trade routes to the towns and cities where the metalworkers had their workshops. This trading helped to spread knowledge about metals and metalwork.

A STATUS SYMBOL

As more metals were discovered and metalworkers' techniques became more skillful, metal possessions became a symbol of power. The large civilizations of the Middle East and eastern Mediterranean had powerful governments and strong armies. Precious metals were fashioned into ornaments and jewelry which showed a person's rank in life. The bronze armor, helmets, and weapons of an army were symbols of its strength.

In Europe, civilizations were less sophisticated, but chieftains still wanted possessions to show their importance. A chieftain might carry a bronze dagger and fasten his cloak with a large ornamental bronze pin or gold buttons. His wife would have gold necklaces, earrings, and broaches.

The Spanish conquistadores found rich hoards of gold treasure when they invaded the Aztec and Inca empires in the sixteenth century. These South American peoples used metals only for ornament, and not for weapons or armor at all. They shaped their work by hammering or by the "lost wax" method. A clay mold was covered with a thin layer of beeswax and another layer of clay. When the mold was heated, the wax melted and flowed out, leaving a gap into which molten metal could be poured.

THE IRON AGE BEGINS

Gold and silver were always to be luxury metals for the rich. Bronze was the dominant metal used for tools until about 500 B.C. In about 2000 B.C., however, a new metal was discovered which was much more plentiful than the others. This new metal was iron.

Iron was probably first smelted in Asia between 2000 and 1500 B.C. but it was not widely used until about 500 B.C., because it was a difficult metal to work. It would only melt at temperatures much higher than early furnaces could produce. Early metalworkers managed to smelt the iron from the rock ore, but it was soft and

◁ *Very ornate jewelry was a mark of someone's rank.*

In this early Chinese blast furnace, four men are operating the bellows to provide air. Molten metal flows out at the bottom of the furnace and runs into a trough. It is stirred to make it stronger before molding. Iron could also run into molds to set into solid blocks known as "pig iron." This could then be remelted for casting or used to make wrought iron.

spongy, not liquid. It could be hammered into shape but not cast in molds. Hammering produced "wrought iron," which was hard and strong because the hammer blows strengthened the metal. But it was not possible to make cast iron.

EASTERN EXPERTISE

In China, however, the story was different. China was a mystery to the West in ancient times and developed quite independently of the other civilizations. As early as the fourth century BC, Chinese metalworkers had discovered a way of heating iron to such temperatures that it melted and could be cast in molds.

First, the Chinese discovered how to make a "blast furnace." In this type of furnace, the flames are fanned by a blast of air to produce greater heat. The early Chinese blast furnaces did this with huge bellows worked by people.

Two other factors helped the Chinese metalworkers. One was the types of clay found in China. The metalworkers found that blast furnaces built from these clays retained the heat very efficiently. They also discovered that they could lower the melting point of iron by mixing it with a mineral known as "black earth."

Cast iron was easier to shape but it was more brittle than wrought iron. The Chinese discovered that annealing the metal, or heating it and then hammering it, made it less brittle. They began to tackle ambitious projects with their new material. They built complete pagodas out of cast iron and, in A.D. 695, the empress of China ordered a 105-foot (32-meter) iron column to be built.

THE WEST CATCHES UP

Meanwhile, things were progressing far more slowly in the West. Early iron-makers used charcoal made from partly burned wood as fuel for their furnaces, and they made an interesting discovery. If carbon from the charcoal mixed with the iron, it made the iron stronger. The Greeks and Romans managed to improve the quality of iron weapons by strengthening the metal with carbon during the heating process. This was the first steel.

The first European blast furnace was developed in Germany in the fifteenth century. These blast furnaces were driven by water and later by steam, and huge bellows fanned the flames. Ironworkers began to produce cast iron but they relied on charcoal as a fuel. As the iron industry grew, supplies of charcoal began to run out.

The breakthrough came in the eighteenth century when an Englishman, Abraham Darby (1678–1717), began to use coke for smelting iron. Coke is made by burning coal in a closed oven to get rid of the sulphur gas in it. The resulting fuel gives off far greater heat than coal or charcoal.

As a young man, Darby was apprenticed to a brewery in Derbyshire. The brewers had discovered that drying malt over a coal fire gave the beer a taste of sulphur, so they experimented and learned how to make coke. In 1709, Darby set up his own iron foundry in Shropshire and began smelting iron using coke in the furnaces. His business was an instant success and other foundries adopted coke as a fuel. It is still the fuel used by the iron and steel industry today.

REDISCOVERING STEEL

The secret of making steel by mixing iron and carbon had died with the Roman Empire. The next important development was to rediscover this secret. This happened in 1740, when Benjamin Huntsman (1704–76) heated a mixture

▷ *Modern metalworkers wear thick protective clothing to protect them from the molten metal.*

THE BESSEMER CONVERTER

Benjamin Huntsman discovered a way of making steel but manufacturers wanted to find ways of producing it more cheaply. In the 1850s, a British inventor, Henry Bessemer (1813–98), developed a new process. Air was blown through molten iron to burn out all the carbon in the metal. The exact amount of carbon needed to make steel was then added to the purified iron. This process was carried out in large container called a "converter." The molten metal was poured into the container and air rushed in at the sides. Another metal, manganese, was then added to absorb the surplus oxygen in the mixture and make the steel stronger. The converter could be tipped on its sides so the molten steel could flow out.

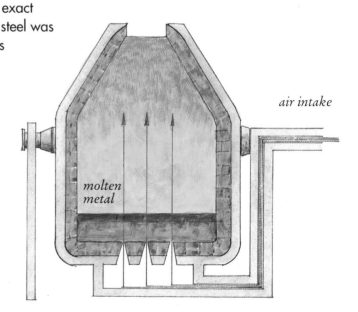

air intake

molten metal

of wrought iron and a carefully measured amount of carbon to produce steel. Steel production took off during the European Industrial Revolution, which began in Britain in about 1750. Its strength and hardness made steel the ideal metal for household items such as cutlery, pots and pans, but it was also used for the large-scale machinery which was needed to get mechanized industry under way.

And so, nearly 9,000 years after the discovery that metal could be melted and shaped, the world was progressing towards the day when to live without it would seem impossible.

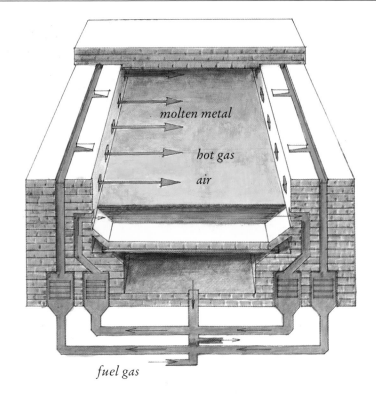

molten metal

hot gas

air

fuel gas

RECYCLING METALS

The fact that metal can be melted down for recycling has always been an important factor. The "open-hearth process" was an early method of melting scrap steel so that it could be used again. It was invented by two German brothers, Ernst (1816–92) and Karl (1823–83) Siemens. Air was heated up as it went into the furnace. This kept the temperature in the furnace high and helped to melt the metal more efficiently.

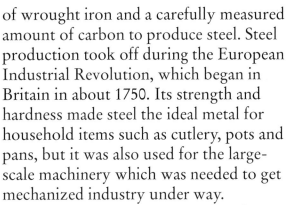

LETTERS & SIGNS

From its beginnings as a simple system for keeping records, writing has developed so that it can be used to record and pass on everything from short messages to the most complex information.

H istory could only truly begin once a system of writing things down had been developed. Before that, people had to rely on word of mouth to pass on information. They could only learn as much as older people remembered about the past. Stories became jumbled and were only shared among members of a family or group, so future generations have had to guess and piece together conflicting pieces of information about this time. Once writing had been developed, the way was open for people to record their thoughts and observations and to detail the history of their civilization, its way of life, customs and religion.

But writing was not invented for these reasons. It began as a means of recording official matters such as taxes and payments for trading goods. Writing was first developed by the ancient Sumerians

△ *Sumerian clay tablet*

▷ *A reed stylus was used to make the wedge-shaped symbols of Sumerian cuneiform writing. The three scripts shown in the foreground are Egyptian, Roman, and Chinese.*

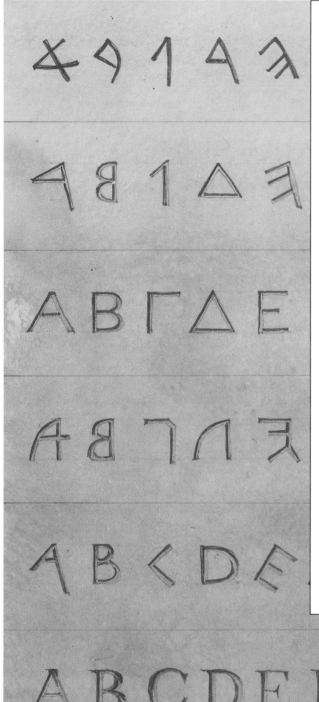

CRACKING THE CODE

When archaeologists first discovered ancient writing such as the hieroglyphs of ancient Egypt, they were baffled by them. What did they mean? How could the code be cracked?

With Egyptian hieroglyphs, the breakthrough came in 1822, when a Frenchman, Jean-François Champollion (1790–1832), solved the mystery of the Rosetta Stone. This slab of stone covered with writing was found by French soldiers in 1799. Champollion realized that some of the symbols were repeated in an ancient Greek script which scholars could already decipher. By matching the hieroglyphs to the Greek letters, he worked out what the writing said.

The Linear B script presented more of a problem than other forms of Greek writing. Its riddle was worked out by a British architect and scholar, Michael Ventris (1922–56). Ventris spent ten years studying clay tablets found at the Palace of Knossos on Crete and at Mycenae on the mainland of Greece. In 1952, he pieced together the evidence and realized that the Minoan Linear B was a primitive form of Greek.

◁ *These examples show how the alphabet developed from the Phoenician script to the modern Roman alphabet we use today. The Romans adopted the Greek alphabet from the Etruscans who ruled over central Italy from 800 to 200 B.C. The Latin script used in the Roman Empire has hardly changed in nearly 2,000 years.*

in Mesopotamia about 5,500 years ago. These early writings were marks scratched onto limestone tablets. In about 3000 B.C., scribes in Mesopotamia began to write on soft clay tablets which were then baked hard in the sun.

At first, the writing itself was in the form of pictures. A simple picture or symbol represented each object. This system was laborious because scribes had to master more than 2,000 symbols for the words they needed. It was also difficult to provide any descriptive information about the objects they were illustrating. So scribes introduced less rigid symbols, known as "ideographs." A single symbol not only represented an object but also ideas connected with it. For example, a circle could mean either the sun, light, warmth, or daylight.

CUNEIFORM

It still took a long time to produce writing by this method because it was difficult to make recognizable drawings with a "stylus," the writing implement used by the scribes. This was made of reed or wood and had a wedge-shaped tip. Gradually, a more abstract system of simple wedge-shaped symbols made with the tip of the stylus developed. This was known as "cuneiform" writing, from the Greek for "wedge-shaped."

The Mesopotamians began to use these symbols to show sounds as well as objects, and to string sounds together to make whole words. For example, if you were putting together the word "belief" in cuneiform, you would use the symbol for "bee" and the symbol for "leaf," even though these words have nothing to do with the word they are spelling. But even with this new system, scribes still had to learn about 600 symbols, so very few people could read or write.

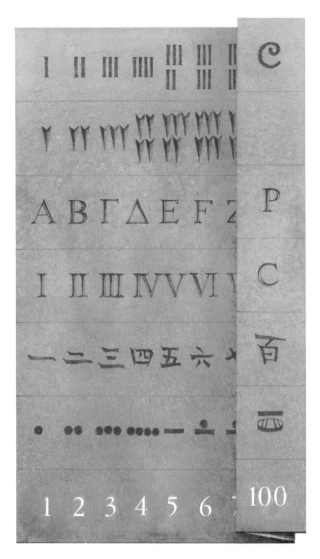

△ The earliest form of counting was to use fingers and thumbs, but as trading progressed, a more formal system was needed to record numbers. The Egyptians and Babylonians used strokes grouped together. The Maya grouped dots. The Greeks adapted alphabet symbols and the Romans used a mixture of strokes and letters. Modern numbers evolved from early Indian numbers.

Cuneiform was also adopted as a writing system by the Babylonians, the Persians, and the Assyrians, although their languages were very different. Most early tablets are tax records and details of ownership and sales, but from about 2400 B.C., people began to use writing in more imaginative ways such as for poetry, letters, and magic spells.

HIEROGLYPHS

Meanwhile, the Egyptians had developed their own style of writing which was used from about 3000 B.C. This is known as "hieroglyphic" script. The writing was mainly used for inscriptions on buildings and tombs. Hieroglyphs were pictures but they could be adapted in different ways. There were about 700 picture symbols which could be used to represent either the object they showed or words which sounded similar. They could also be put together as the syllables of a longer word. Twenty-four extra signs represented single consonant sounds, to write words that could not be illustrated.

Egyptian scribes gradually introduced a quicker, simpler writing called "hieratic," although tomb inscriptions were still written in hieroglyphs.

Two civilizations in Mexico invented their own form of picture writing. The Maya, who lived on the Yucatán peninsula from A.D. 300 to 900, carved their picture symbols, or "glyphs," onto huge stone pillars and also painted them on to long strips of bark paper which were folded into books. The Aztecs, who settled in Mexico in about 1325, used glyphs which were probably developed from the Mayan system, although they are easier to decipher.

WRITING SOUNDS

All these picture symbols could be written in any direction, which made them even harder to understand. In Egypt, the reader was helped by the fact that the characters faced the way in which the pictures were meant to be read.

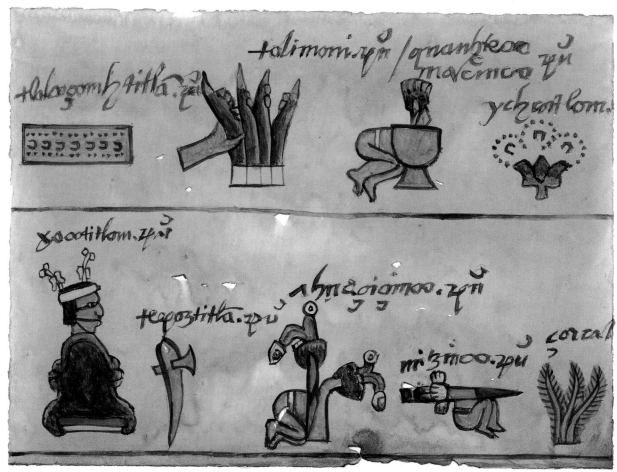

The Aztecs of Mexico used a sophisticated picture script. Spanish translations are written in alongside.

△ *A selection of scripts, from left to right: ancient Egyptian hieroglyphs; the Minoan Phaistos disk which has not yet been deciphered; Greek numerals; Roman letters; Arabic script; Chinese characters; Maya glyphs.*

These early forms of writing were difficult to master because there was a different symbol for every object or every syllable of a word. This meant that there could be thousands of symbols to memorize, and it was beyond the skill of most people to learn how to read and write.

The big breakthrough in the history of writing came when people began to realize that all the syllables were made up of the same few sounds, and that each of these could be shown by a single symbol or letter. This discovery was made in about 1600 B.C. by the peoples who lived around the shores of the eastern Mediterranean. Their alphabets are known as "Semitic" scripts. The Canaanites, who lived in what is now Israel and Lebanon, developed an alphabet in which each symbol was named after the object it was based on. For example, the Canaanite word for "ox" began with an "a" sound, so the "a" sound was represented by the head of an ox. The Phoenicians, who lived along the shores of the Mediterranean in Syria and Lebanon, adapted the Canaanite script for their alphabet. They were a seafaring people, and their alphabet spread far and wide. But it contained only consonants, so it was more like shorthand than a true alphabet.

WHICH WAY TO WRITE?

The Greeks adapted the Phoenician script but introduced vowels as well as consonants. New Greek scripts evolved as time went by. Two of these, which are known as "Linear A" and "Linear B," were used by the Minoans who lived on the island of Crete. These scripts were the first to always be written in horizontal lines. At first, the Greeks wrote from right to left. Later, they wrote from right to left and from left to right on alternate lines, so the reader zigzagged back and forth. This style is known as "boustrophedon" which means "in the way an ox plows a field." Finally, they settled on the system of writing from left to right, as we do today.

AN UNSOLVED MYSTERY

The Greek Linear B script developed from an earlier form known as Linear A. This script is only found in about 400 inscriptions on clay tablets. In spite of numerous attempts to decipher it, Linear A remains an unsolved mystery.

The Chinese wrote their characters with a brush and a cake of ink. The name written here says K'ung-Fu-tzu. This man is better known in Europe as Confucius (c. 551– 479 B.C.), the great Chinese philosopher.

THE ALPHABET

The word "alphabet" comes from "alpha" and "beta" which are the first two letters of the Greek alphabet. If you look at the classical Greek alphabet, you can see many similarities with the alphabet we use today. This is because the Roman alphabet used in many parts of the world was developed from the later version of the Greek alphabet.

By this time, the Greeks had also adapted the way they wrote letters to suit the materials they were using. If letters were to be carved in stone, it was easier to have straight lines so the letters were pointed with sharp angles. For writing on papyrus or parchment, a more rounded style flowed better. The Romans followed this through, changing angular Greek letters to rounded ones and also dropping some of the letters altogether. By the seventh century B.C., the Roman alphabet had twenty-one letters. By about the first century B.C., the letters Y and Z were added, and finally J, U, and W were added in the Middle Ages. With the simplified alphabets of Greece and Rome, writing was within the grasp of everyone.

A different alphabet evolved in the East, although it, too, stemmed from the Semitic scripts of the eastern Mediterranean. This was "Aramaic" and it probably first appeared in the tenth century B.C. It is the ancestor of both Hebrew and Arabic scripts. Arabic is the language and script of the Koran, the holy book of the Islamic religion, so this script spread through followers of this faith. The modern 28-letter Arabic alphabet is written from right to left.

ONE SCRIPT, MANY TONGUES

One form of writing stayed separate from the rest. This was Chinese, which does not have an alphabet but consists of thousands of symbols, or "characters."

Unlike other scripts, the Chinese system has become more complicated with time. During the Shang period (about 1766–1122 B.C.), there were about 2,500 signs. Today, there are about 50,000. Some of these are pictures of objects, some suggest abstract ideas. The writing is difficult to learn but it has an advantage in China, where many languages are spoken. With other scripts, if you see writing in a foreign language, even though you recognize the alphabet, you cannot understand the words unless you know that language. But, as Chinese writing represents a word with symbols instead of spelling it out, people do not have to speak the same language to understand it. In 1979, the Chinese introduced "pinyin," an alphabet of fifty-eight letters used for writing proper names and place names.

The Chinese style of writing spread throughout the Far East. Japan adopted it in the third century AD, but it was not ideally suited to their language. Five hundred years later, they developed a script which was broken down into syllables. This new script meant that fewer symbols had to be memorized. Today, there are two styles of writing in Japan, one for official documents and one for literature. Each has only fifty symbols.

And so, through people's need to record details of trading, religious beliefs, and their thoughts and imaginative ideas in the form of literature, the concept of writing spread throughout the world.

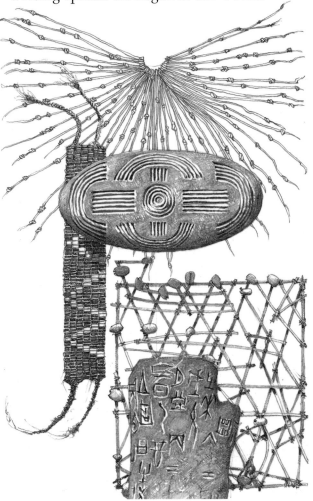

△ *Some societies did not use writing to communicate. The Incas of Peru used the "quipu," in which different colored cords were knotted in various ways to record information. The Iroquois Indians of North America used patterns and colors on their wampum belts to pass on information. The Aborigines of Australia used message sticks in which marks or grooves were cut. In medieval England, sticks with notches cut in them were still used to record financial matters, even though written scripts were widespread by then.*

THE BOOK

Today, we can find books on almost any subject, but it took centuries to arrive at a quick, cheap way of printing them. The first books were so precious that only a few people were ever allowed to touch them.

Our world is full of books, magazines, and all sorts of printed materials. We can read to learn or for pleasure. We can find books to help us with everyday living or show us the beauty of the world around us. There can be no subject in the world that people have not written about at some time. But in the days of the early civilizations, it was a very different story. The book developed in stages over thousands of years before the invention of the printing press made books available to everyone.

Early clay tablets were not a very practical material for producing books. They were unwieldy to read and storage was a problem. Even so, a few ancient writings have been found on clay tablets. The oldest known written poem, the *Epic of Gilgamesh*, was found on twelve clay tablets in the library of King Assurbanipal of Assyria (669–627 B.C.), although the poem was written long before his reign. Gilgamesh was a legendary king of Uruk in Mesopotamia.

△ *Records written on papyrus scrolls could be stored for a long time in Egypt's dry climate.*

The poem tells of his adventures trying to find the secret of eternal life. The tablets were written in Akkadian script, a type of cuneiform writing which was developed after King Sargon of Akkad (c. 2334–2279 B.C.) came to the Assyrian throne in 2300 B.C.

When archaeologists excavated the Royal Palace at Ebla in the Near East, they discovered a library of 1,900 clay tablets dating back to about 2000 B.C. These tablets record over 140 years of Ebla's history. Findings of this kind are rare, however. Before writers could let their imaginations have free reign in the creation of literature, better writing materials had to be introduced.

King Assurbanipal of Assyria.

PORTABLE PAPYRUS

The first stage was to find a less cumbersome surface to write on than clay tablets. In about 3500 B.C., the ancient Egyptians discovered that the papyrus reed which grew by the river Nile could be made into a type of paper. The reed grows to ten feet (three meters) tall, and has thick triangular stems filled with a white, spongy pith.

The Egyptians removed the green outer skin and cut the pith into thin strips of equal length. They laid the strips side by side, with their edges overlapping, to make a sheet of the size they wanted. Then they laid another layer of strips at right-angles to the first layer. The layers were dampened and pressed under a heavy weight, or hammered flat. As the fibers dried, they knitted together to form a thin sheet of writing material also known as "papyrus," which gives us the English word "paper." Scribes wrote on papyrus with a reed pen dipped in ink. The earliest types of ink were made from water mixed with vegetable gum and soot or vegetable dye.

▽ *Books have long been used to spread all kinds of knowledge, from the scientific discoveries of ancient Greece to the religious writings of the Middle Ages. When printing was invented, more books were written for entertainment, such as poetry and stories.*

Early books were not made of sheets bound together as they are today. The first books came in the form of clay tablets, and later books were in the form of rolls of papyrus, silk or parchment.

Papyrus soon became the most common type of writing material in Egypt and continued to be used for thousands of years. It was exported from Egypt to other Mediterranean lands, and later civilizations such as the Greeks and the Romans made their own papyrus from imported reeds.

The disadvantage of using papyrus was that the papyrus reed only grew wild in Egypt, so everyone had to rely on only one source of writing material. If Egypt stopped supplying it for some reason, then scribes throughout much of the world would have nothing to write on. So a new type of material was developed.

A NEW MATERIAL

"Parchment" was made from sheep, goat or calf skin. A more delicate version, "vellum," was made from the skins of very young animals. Eumenes, King of Pergamun in Asia Minor (197–159 B.C.) is said to have invented parchment in the second century B.C. Legend has it that the pharaoh of Egypt stopped supplying papyrus because he was jealous of the library at Pergamun and wanted to stop new books being added to it. Refusing to be thwarted, Eumenes turned his mind to finding a new material for his scribes to write on. The name "parchment" is said to come from the word "Pergamun."

Parchment and vellum were made by soaking the animal skins in lime, and then stretching them and scraping them clean. The skin was rubbed with a pumice stone until it was perfectly smooth.

Parchment was the main writing material until well into the Middle Ages. People wrote on it with a quill pen made from a goose feather. This combination was so successful that medieval monks could produce exquisite illuminated manuscripts in which letters were decorated with designs in gold, silver and brilliant colors. The sheets of parchment were sewn together to make books protected by covers made from leather or wood. And so the book as we know it began to appear.

PAPERMAKING

In the East, writing materials were developing in different ways. In China, people wrote with the same type of brush that they used for painting. In fact, writing and painting were often combined, particularly in poetry. Early books were produced on rolls of silk or strips of bamboo. But the Chinese were also responsible for one of the most important developments in the history of the book. This was the invention of paper.

Paper was first made in China in about A.D. 50. Old fishing nets, hemp and rags were beaten in water until they were a pulp of fibers. The pulp was spread onto a bamboo screen. The water drained through the screen, leaving a mat of fibers. Then, all the water was pressed out of the fibers and the paper was dried in the sun.

Medieval books were decorated with elaborate colored or "illuminated" initials.

It was several hundred years before the secret of papermaking reached the West, and even then it happened quite by chance. The city of Samarkand in Uzbekistan was the main junction of the Silk Road, the ancient trading route between China and the Mediterranean. During the Siege of Samarkand in A.D. 768, Arabs conquered the city and captured many Chinese prisoners of war. Among these prisoners were papermakers who passed on the secret of their craft.

The Arabs began to develop their own papermaking industry which spread from Samarkand to other cities such as Baghdad, Damascus and Cairo. Papermaking skills reached Byzantium in the eleventh century and finally spread to western Europe in the twelfth century.

The development of books grew with the spread of papermaking. In A.D. 900, Baghdad had about a hundred workshops where scribes copied books for sale. Centers of learning such as Toledo in Spain and Fez in Morocco were also papermaking centers where books were produced.

PRINTING BLOCKS

Copying books by hand was a laborious business and, once again, it was in the East that the next development took place. During the ninth century, printed books began to appear in China. The text and pictures were carved on wooden blocks which were coated with ink. The blocks were then pressed on to paper to print the page. The Chinese also developed single pieces of type which could be moved around in different arrangements, but this was a very complicated system to use for Chinese writing, because separate blocks were needed for each of the thousands of different Chinese characters.

RELIGIOUS BOOKLETS

The first printed books in Europe appeared in monasteries, where monks had been copying by hand for centuries. The monks used carved woodblocks to print the books. At first, blocks were simply made to print decorated letters which appeared often. Then, the monks began to print religious pictures with short texts and then short booklets. The monks printed these in the language of the country they were in instead of in Latin, the official language of the Roman Catholic church that only priests and scholars could understand. Suddenly it was possible to spread information to a wider audience.

The woodblock method was satisfactory for printing short works but it had disadvantages for longer books. A separate block had to be carved for each page, which took a long time, even when the work was done by skilled woodcarvers. Each block

▽ *Medieval monks spent much of their time copying books by hand. One book could take hundreds of hours of work.*

WRITING THE BIBLE

The way the Bible came to be written is a good example of how early books were first written. The earliest versions of the Old Testament were written on scrolls of parchment. Many people contributed to the Old Testament and although most of it was written in Hebrew, two of the books were written in Aramaic. Scribes had to keep making new copies because the scrolls disintegrated as time went by. The Dead Sea Scrolls, found in a cave by a shepherd boy in 1947, are more than 2,000 years old and are the oldest copies of the Old Testament found so far.

The New Testament was written in Greek. The earliest books were written on scrolls of papyrus, but in later versions the paper was folded to make a "codex."

In Roman times, the Bible was written in Latin. Then, in the fourteenth century, scholars began to translate the Bible into other languages. In England, a scholar called John Wycliffe (c. 1329–84) made the first English translation.

Many English versions appeared after that, but the old-fashioned English of these older Bibles can be difficult for people to understand today. Modern versions rewrite the Bible in everyday language.

could only be used to print one page of one book, so a lot of time was wasted in carving blocks that would only be used for a small number of copies of a booklet. It was also difficult to carve the letters clearly enough to make a good impression on the page. Sometimes the wood warped which made the carving useless. So, before whole books could be produced, printers had to look for a more efficient method. The answer was to use metal instead of wood.

MOVABLE TYPE

In the middle of the fifteenth century, metalworkers in different parts of Europe began to experiment with making metal type. The printing press was invented in about 1450 by Johannes Gutenberg (c. 1400–1468), who was born in the German town of Mainz. He was a skilled goldsmith, but he was also fascinated by books. He would spend hours in the monastery library, studying the books and watching the monks copying them out by hand. He realized what a slow process this was, and wondered if metalwork could be used to produce the books more quickly. The idea came to him that the letters of the alphabet could be cut on separate blocks of metal, called "movable type," which could be moved round to form words.

In 1428, Gutenberg moved to Strasbourg and set up in business with a partner, Johann Fust (1400–66), who was to put up the money for the equipment they needed. In 1456, they produced the first book printed with movable type, the famous Gutenberg Bible. But Fust wanted the business for himself and so he demanded his money back from Gutenberg. Gutenberg could not pay and was forced out of the business he had invented. In 1457, Fust printed a Psalter,

The first printed books did not look very different from handwritten ones. The letters were shaped in the same handwritten style, known as Gothic script, and the pictures were printed from woodblocks and colored by hand.

or Book of Psalms, which was mainly Gutenberg's work.

After Gutenberg's invention, however, books became more widely available and printing industries were set up in many cities, including London, Paris, Rome, Venice, Florence, Milan and Cologne. Gutenberg had printed his book in Gothic type, which was an imitation of the handwriting used by the monks. By 1470, printers were using Roman type, with separate, upright letters, far easier to read than Gothic. The rounded, sloping letters of Italic type appeared in 1495.

SPREADING IDEAS

By this time, a new period of learning had begun. The Middle Ages were followed by the Renaissance or "rebirth" of Europe. From the middle of the fourteenth century, there was new interest in the works of the ancient Greeks and Romans. The Renaissance began in Italy, where rich and powerful people such as the Medici family encouraged the work of painters, sculptors, poets and scholars.

ASKING QUESTIONS

People became hungry for more knowledge and began to question the ideas of the past. They also began to question the behavior of the all-powerful Roman Catholic church which had been dominant for 1,000 years. This new awakening was helped by the invention of printing.

By 1500, there were printing presses in 143 European towns and more than 16,000 different works had been published. Counting all the different editions, such as separate editions of the Bible produced by different printers, there were 40,000 printed titles. At first, most books were Bibles, prayer books and Psalters, but before long other books began to appear, such as translations of ancient Greek and Roman works and law and science books. Influential thinkers such as the Dutch scholar Erasmus (1466–1536) could circulate their work to a wide audience through printing. Erasmus's work *In Praise of Folly*, published in 1509, attacked the Catholic church for squeezing money out of its followers to line its own pockets.

▽ *Gutenberg's print shop was crowded and busy. On the right, the first printer is screwing the heavy press down on the paper which lies on top of the bed of type. The printer in the background is coating the type with ink. An ink dabber is shown in the center foreground. On the left, letters are being removed from the wooden storage case and assembled in a line in a narrow frame known as the composing stick.*

Criticism of the church led to another Renaissance movement, the Reformation. Rebellion against the domination of the Roman Catholic church spread across Europe from about 1500. It was fired by Martin Luther (1483–1546), who was born in Saxony. Luther, who had studied religion and become an Augustinian friar, went to Rome on a pilgrimage and was disgusted by what he saw. He told of priests who sold "indulgences," or pardons for sins, and so-called religious relics which were no more than animal bones.

Printing helped his cause because he was able to publish pamphlets by the thousands. These pamphlets, urging people to rebel against the church, were circulated throughout Europe and helped to end the dominance of the Roman Catholic church.

In England, the first printing press was set up by William Caxton (c. 1422–91), in 1476. Caxton had learned the printing trade in Cologne and he produced about eighty titles at his printing press in London. England did not have the painters and sculptors of Italy, but it was famous for its playwrights and poets such as William Shakespeare (1564–1616) and Edmund Spenser (c. 1552–99). The first edition of Shakespeare's plays was published in 1623.

POWER-DRIVEN PRESSES

In these early presses, the letters were assembled by hand. The problem was to keep the letters firmly in place while several copies were printed. After experimenting with various methods, printers came up with the idea of arranging the letters in a wooden frame called a "forme." The paper was pressed on to the letters from above. Everything

was set up and operated by hand.

Faster printing became possible in the nineteenth century when power-driven presses were introduced. In the cylinder press, a cylinder rolled the paper over a flat surface which held the type. In the later rotary press, the type was also on a cylinder. In 1884, the Linotype typesetting system was invented. Instead of positioning each letter separately, the typesetter typed words for a whole line of type using a keyboard. A solid line of type was then formed from molten metal.

MAKING IT LEVEL

Applying the ink was also a problem with the early presses. At first, ink was applied to the metal type with two circular "dabbers." The printer covered them with ink and then dabbed ink onto the type. It was difficult to coat the type thoroughly once it was arranged in the

△ *By the nineteenth century, printers used these enormous presses. As more people learned to read, books and newspapers flourished.*

press, so the bed on which the type was laid was put onto a sliding carriage. The type could be slid out to be inked and then pushed back into the press.

If each piece of type was not perfectly level, some letters would not print properly. For example, if a letter was even slightly lower than the others, the paper would not come into contact with it when the press was applied. The answer was to put several sheets of paper under the sheet being printed. This allowed enough "give" to push to paper on to each letter when pressure was applied.

Printing has continued to develop and become more sophisticated but nothing achieved since has had the same impact as the invention of movable type, which opened the world of books to everyone.

SEEING CLEARLY

The discovery of lenses not only helped to change the lives of people with long or short sight. It has allowed us to examine the tiniest living things and to see far into the universe.

Throughout history, there have always been people who suffered from poor eyesight. Some people are shortsighted, meaning they cannot see things unless they are very close. Other people are longsighted, which means that they can see into the distance but cannot see things close to them. As people get older, their eyes often do not work as well, and they may not be able to see clearly to read, drive or watch television.

Today, poor eyesight can be corrected with spectacles containing lenses made from glass or plastic. Many people with short or long sight wear contact lenses instead of glasses. These fit straight onto the eyeball and do not show at all.

The discovery of lenses did not just improve people's everyday sight. It also led to the invention of other optical instruments, such as microscopes and telescopes. Today, powerful microscopes allow scientists to see things invisible to the naked eye, and huge telescopes give a clear picture of distant stars and planets.

△ *It is said that Emperor Nero improved his vision by looking through a jewel.*

Early glass was so full of impurities that it was impossible to see through it. Glass in windows could let light in, but was not much use for looking through. Even when glassmakers began to make glass clear enough to see through, they could not make it flat. Curves in the glass made things look distorted which was not very satisfactory for windows, but probably helped in the invention of lenses. People must have noticed that when glass curved in certain ways, it made things look bigger or smaller.

EARLY LENSES

To develop efficient lenses, people had to understand how light behaves. Light rays bend or "refract" when they pass from one material to another. When they strike a surface, some of the light is reflected. Light rays are reflected to your eyes from everything around you. The lens in the eye bends the light to focus an image of these objects on the back of the eye. If the lens does not focus accurately, the image is blurred. This can be corrected with a glass lens in front of the eye which bends the light to give a sharp image.

Lenses dating from about 2000 B.C. have been found in Crete and western Asia, but they were made of poor glass and would have been useless as optical tools. Over 1,500 years passed before the Greek scientist Euclid (c. 300 B.C.) wrote about the refraction and reflection of light. It is said that Nero (A.D. 37–68), Roman emperor from A.D. 54 to 68, watched gladiators fight through a jewel held to his eye. If this is true, he must have been shortsighted, and the jewel must have acted as a lens to focus on distant objects.

HOW LENSES WORK

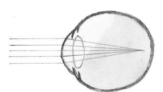 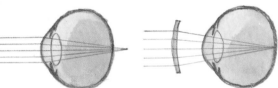

short sight
retina

concave lens (first appeared 1568)
image focused behind the retina

long sight
retina

convex lens (first appeared 1280)
image focused in front of the retina

Lenses were probably discovered by accident when people noticed the effects caused by curved pieces of glass. They began to make lenses by casting disks of glass in a mold and then grinding and polishing the glass until they had the smooth, curved surface they wanted. At first, this work was done by hand, but then lens-makers used a lathe to grind the glass. A later discovery was that it was better to cut a lens out of a block of glass instead of casting the glass in a mold. This technique was developed during the late seventeenth and early eighteenth centuries.

Scientists may have been working on improving lenses at this time but the main developments seem to have begun in the Middle Ages, when scientists in the Islamic world were developing lenses. The English writer Robert Grosseteste (c. 1175–1253) also experimented with the effects of mirrors and lenses and foretold of a day when lenses would be strong enough to show things that were not visible to the naked eye.

Anton van Leeuwenhoek

Grossteste's experiments inspired the English monk, scholar, and scientist Roger Bacon (c. 1214–92). In his writings, Bacon prophesied the invention of microscopes, telescopes, airplanes and steam engines, and it is also thought that he invented the magnifying glass.

Spectacles were being made at about the same time. In 1280, a scientist named Salvino degli Armati (1245–1317), from Florence, Italy, made two eyeglasses which magnified. At first, Armati tried to keep his discovery a secret, but word soon got round. By the late Middle Ages, spectacles were quite common.

THE MICROSCOPE

Much of the work on developing lenses went on in Holland. Dutch spectacle-makers continued to improve the lenses they made, and, by the late fifteenth and early sixteenth centuries, they were making lenses powerful enough to use in simple microscopes.

One of the most important pioneers of the microscope was the Dutch scientist Anton van Leeuwenhoek (1632–1723). Leeuwenhoek made lenses with stronger magnification than any that had been made before. The most powerful magnified everything to about 300 times its actual size.

Leeuwenhoek used his lenses to study living organisms. He discovered bacteria and observed the composition of muscles and red blood cells. He also studied the life cycle of various animals, including the flea, the ant, and the weevil.

GREATER MAGNIFICATION

At the same time, other scientists were developing another type of microscope, the "compound microscope." This uses two or more lenses to enlarge the image. Greater magnification was possible with compound lenses, but it was difficult to get a clear image.

We are not sure who invented the compound microscope because several scientists were working along the same lines. It was probably invented at the end of the sixteenth century by a Dutch spectacle-maker, Zacharias Janssen (1580–1648). Janssen's microscope was described in 1609 by the Italian scientist

▷ *Some important early optical inventions:*

1 Anton van Leeuwenhoek's microscope had a double convex lens mounted between brass plates.

2 The compound microscope looks more like the microscopes of today. It had two lenses held in an adjustable tube.

3 Robert Hooke's microscope used a lens to focus light on the subject. The writer in the background wears a pair of spectacles to help him see his work.

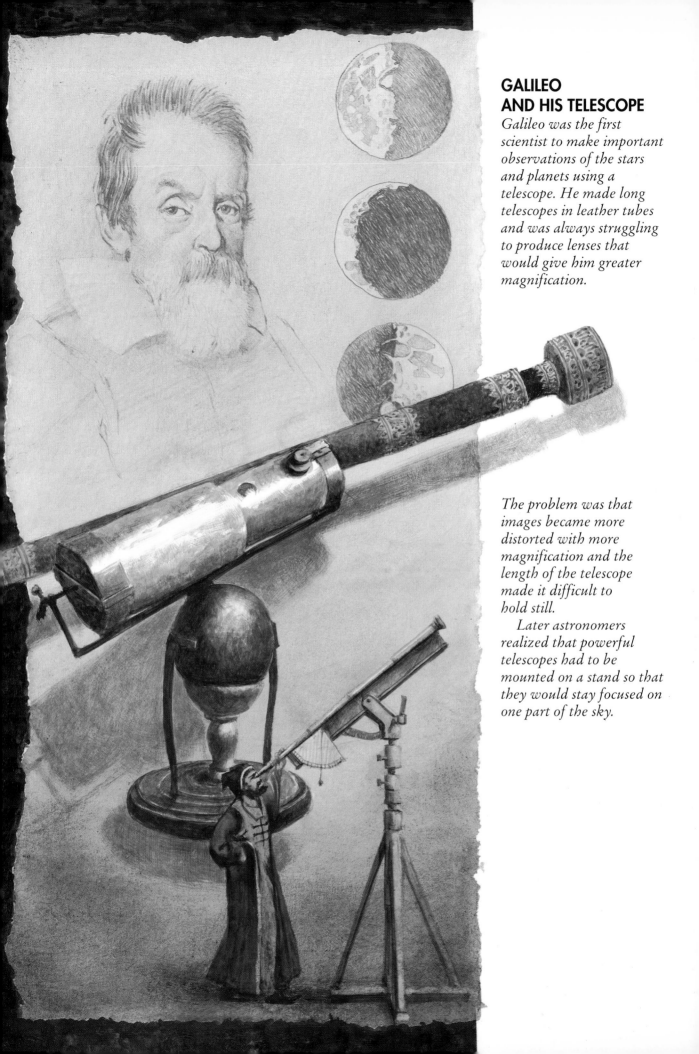

GALILEO AND HIS TELESCOPE

Galileo was the first scientist to make important observations of the stars and planets using a telescope. He made long telescopes in leather tubes and was always struggling to produce lenses that would give him greater magnification.

The problem was that images became more distorted with more magnification and the length of the telescope made it difficult to hold still.

Later astronomers realized that powerful telescopes had to be mounted on a stand so that they would stay focused on one part of the sky.

Galileo Galilei (1564–1642). It is also possible that the compound microscope was invented by a Dutch lens grinder, Hans Lippershey (died c. 1619). By the middle of the seventeenth century, compound microscopes were being made in Italy, England and the Netherlands.

The first detailed description of the compound microscope was given by the British scientist Robert Hooke (1635–1703). Hooke developed his own microscope and described his work with it in his book, *Micrographia*, published in 1665. Hooke was one of the first scientists to examine vegetable material under the microscope, and in 1667 he discovered that cork is made up of cells.

STUDYING THE STARS

People have been studying the stars and planets since civilization began. Astronomers mapped the stars and noticed that they were in groups, or "constellations," which formed patterns in the sky. They gave these constellations names, often after heroes of mythology. Astronomers in ancient China charted a calendar from the position of the stars at different times of the year.

Some of these early astronomers made important discoveries. Astronomers from ancient Greece put forward particularly impressive theories considering that they had very little equipment to help them. Aristarchus of Samos (c. 310–230 B.C.) worked out that the Earth rotates on its axis and moves round the sun. Hipparchus (c. 190–120 B.C.) plotted the first accurate map of over 1,000 stars and also made an accurate measurement of the distance from the Earth to the moon. Ptolemy (second century A.D.), a Greek astronomer who lived in Alexandria, contradicted Aristarchus by putting forward the theory that the Earth was a

ALL THE COLORS OF THE RAINBOW

Sir Isaac Newton made the discovery that white light is made up of different colors. He proved this by directing a beam of sunlight through a triangular glass block called a "prism." The light rays bent and split into the seven colors of the spectrum: red, orange, yellow, green, blue, indigo and violet. Each color bends a slightly different amount, which is why you see them separately.

The same thing happens when the sun shines during a fall of rain. The raindrops act as prisms and split the sunlight into the colors of the spectrum. You see a rainbow across the sky.

Newton's work with light and optics led him to develop the reflecting telescope in an attempt to cut out the distortion of colors found in earlier simple telescopes.

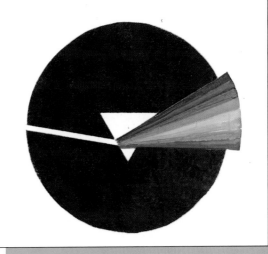

stationary object at the center of the universe and that the sun, moon and planets moved around it. This theory was accepted for over 1,000 years.

THE TELESCOPE

These early astronomers had to view the skies with the naked eye and so their knowledge was limited. For example, our own galaxy of stars, the Milky Way, contains millions of stars but only a few thousand can be seen with the naked eye. It was not until the invention of the telescope that people could really begin to see into space.

The telescope was invented by Hans Lippershey, who had also worked on the compound microscope. It is said that Lippershey held two lenses up to the sky and noticed the magnified image they produced. He made his telescope and patented it in 1608.

SATELLITES AND SUNSPOTS

His main rival was Galileo who made his own telescope in 1609, after hearing of Lippershey's discoveries. Galileo used his telescope to study the sky. He spent the next few years building more powerful telescopes and made many significant discoveries. He discovered that the moon does not have any light of its own but shines with light reflected from the sun. He saw the satellites orbiting Jupiter, and noted the movement of Venus. He found that the Milky Way was made up of many distant stars, and he saw the spots on the sun. He kept a record of these spots over a period of time and came to the conclusion that the Earth moves around the sun, as the Polish astronomer Nicolaus Copernicus (1473–1543) had thought. This disputed the ideas of the ancient Greek astronomer Ptolemy which people had believed for centuries.

Copernicus had made himself very unpopular with this suggestion. The Christian church preferred Ptolemy's theory because it placed the Earth and the people on it in the position of greatest importance. Now Galileo's observations seemed to show that Copernicus was right. However, the Roman Catholic church was not prepared to accept

Galileo's idea. It was fighting for survival against the Protestant reformers at the time and thought a theory which went against all its teachings would only add to people's disillusionment. In 1633, Galileo was forced to take back his theory and say that he had been mistaken. He was imprisoned in his house in Florence where he stayed for the rest of his life. Yet it is said that, as he publicly renounced his theory that the Earth moved round the sun, he was heard to mutter to himself: "And yet it moves."

During the next few years, scientists continued to develop the telescope. The Dutch scientist Christian Huygens (1629–95) made telescopes so long and heavy that they had to be mounted on a

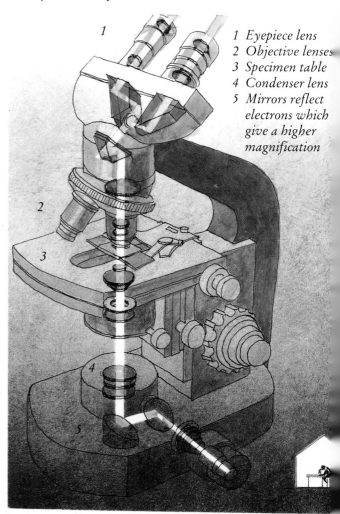

1 Eyepiece lens
2 Objective lenses
3 Specimen table
4 Condenser lens
5 Mirrors reflect electrons which give a higher magnification

△ A modern electron microscope.

stand to steady them. Huygens discovered Saturn's rings in 1656.

THE REFLECTING TELESCOPE

But even though telescopes were becoming more powerful, colors were still distorted and images blurred. This problem was solved by one of the greatest scientists of all, Sir Isaac Newton (1642–1727), who invented a reflecting telescope. Newton's telescope used mirrors. Light fell onto a concave mirror, from which it was reflected onto a second mirror. This mirror reflected the light onto the eyepiece which magnified the image. The image was much sharper than with earlier telescopes, and there was no color distortion.

A British astronomer, William Herschel (1738–1822), developed Newton's ideas and built the largest reflecting telescope then known. In 1781, he discovered the planet Uranus. Mercury, Venus, Mars, Jupiter, and Saturn are visible with the naked eye and were known to the ancients. Neptune and Pluto were discovered later.

Huge reflecting telescopes are still used to view space today. The world's largest telescope is the Russian 21-foot (6-meter) reflector which was set up in the Caucasus Mountains in 1976. Radio and X-ray telescopes are yet more accurate, recording information about objects in space that are much too far away to be seen with an optical telescope.

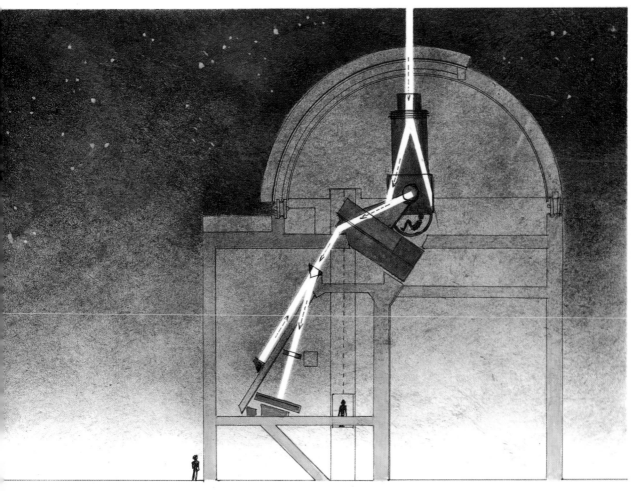

△ *Astronomical telescopes catch light from the stars and planets in a mirror which concentrates the image onto another mirror and so to an eyepiece or screen where the astronomer can view it.*

VOICES ACROSS THE WORLD

Communication links between people have always been important, but we have moved a long way from the beacons and smoke signals used by early people to the fax machines and satellites we have today.

People have always needed to communicate with each other across long distances. Bonfires on hilltops were an early method of signaling danger. In ancient China, bonfires were lit along the Great Wall to warn of attacks from barbarians. The North American Indians used smoke signals. The Romans flashed messages with mirrors turned to catch the sun. Flashing lights and flags have been used in a similar way.

Sending messages in this way was all right in certain circumstances. But the people receiving the signals had to be able to see them, so they had to be sent from a prominent viewpoint which was not too far away. If the messages were complicated, a code had to be used, which had to be understood by everyone. Such signals were usually only used to send messages in times of war, but until the nineteenth century, the only other way of communicating was to send a written message, which took time if the people involved lived any distance apart.

△ *This is the very first telephone, which Bell used to speak to Tom Watson on March 6, 1876.*

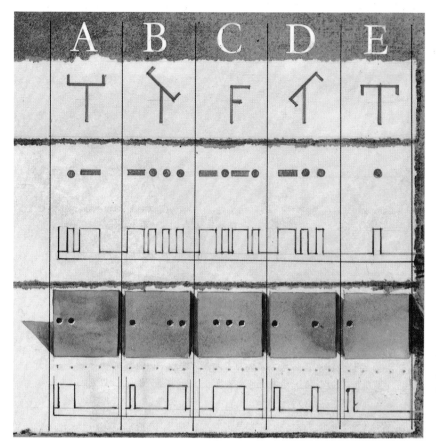

◁ *Semaphore code*
Semaphore, developed in 1794, used a system of moving arms worked by ropes to create symbols for each letter.

◁ *Morse code*
Samuel Morse's code could be transmitted along a wire using a key, shown on the right. The code is shown as dots and dashes and (below) as electrical signals.

◁ *Five-unit code*
This code was developed from the Morse code for using with a teleprinter, an instrument for typing telegraphs to be sent along telephone wires.

THE TELEGRAPH

Then, in the nineteenth century, a new discovery at last brought more efficient methods of communication.

In 1831, British scientist Michael Faraday (1791–1867) made the first electric generator. He found that moving a loop of wire over a magnet produced an electric current. He then tried moving the magnet instead of the wire and found that an electric current was produced again.

Once the link between electricity and magnetism had been established, it could be used in other ways. The first telegraph was patented by British scientist Sir Charles Wheatstone (1802–75) and an Indian Army officer, Sir William Cooke (1806–79), in 1837. It used magnetic needles which pointed at different letters in response to electric currents.

Another type of electric telegraph was invented by an American inventor, Samuel

After the invention of the telegraph, telegraph wires spread all over the land, linking towns and cities, and changing the look of the countryside.

△ *Alexander Graham Bell uses his early telephone equipment. Behind are some of his rough sketches for his telephone.*

Morse (1791–1872). He devised "Morse code," a system which could be tapped out on an electric key. The code for each letter was made up of a different combination of long buzzes or "dashes," and short buzzes or "dots." The message was received by another machine at the other end of the telegraph line, where an operator decoded it. Morse code could also be transmitted with flashing lights. Morse established the first telegraph line between Washington and Baltimore in 1844.

At last, there was an efficient way of sending messages quickly. By the 1860s, telegraph wires connected the East and West Coast of the United States and there was a cable across the Atlantic to Europe.

TRANSMITTING VOICES

The telegraph was a big breakthrough, but it still had disadvantages. The sender and the receiver had to understand Morse code, so messages had to be sent to telegraph offices where skilled operators decoded them. Then the message had to be delivered to the person concerned. It was quicker than sending a letter, but not as quick as being able to speak to the person directly. If coded messages could be sent along electric wires, could the human voice also be transmitted? It seemed a possibility to one man at least, Alexander Graham Bell (1847–1922).

Bell was born in Scotland where he

lived until he was twenty-three. His mother was deaf, and his father specialized in teaching deaf children. Bell also became a teacher of the deaf, and when the family emigrated to Canada in 1870, he continued his work there. He moved to Boston, Massachusetts, three years later. His work with deaf children made him think about speech sounds and question whether they could be transmitted along electric wires in the same way as telegraphs, which by then were well established.

On early telephones, such as Bell's "box" telephone, the earpiece and mouthpiece were combined. There was no dial on these early phones. You had to call the operator to put your call through.

was adjusting the transmitter in another room. Bell continued working on his calculations and, on March 6, 1876, he transmitted the first words, "Come here, Watson, I want you." Tom Watson, listening on the receiver in another part of the house, heard him. The telephone had been invented.

Bell patented his invention in 1876, and in the same year he made the first long-distance call between Brantford and Paris in Ontario, Canada, a distance of 70 miles (110 kilometers). Now, it was only a matter of time before it would be possible to communicate by telephone to anywhere in the world.

COME HERE, WATSON!

Bell knew that sounds make vibrations on the eardrum which the brain translates to make sense of them. His idea was to make a transmitter with a disc which would vibrate when struck by sound waves, in the same way as the eardrum. Sound vibrations from the transmitter would pass along a wire to a receiver which would also have a vibrating disc. This receiver would convert the sound vibrations back into words.

Bell spent two years making his transmitter. The breakthrough came in 1875 when Bell, listening on the receiver in one room, heard distinct sounds from his assistant, Watson, who

1920s candlestick telephone.

RIVALS

But Bell was not the only person who had come up with this idea. An American, Elisha Gray (1835–1901), was also developing a telephone quite independently of Bell. Gray patented his invention on the same day as Bell. A long court case followed to establish who had actually invented the telephone. The judges finally decided in Bell's favor because he had filed his patent at noon, and Gray had not filed his until two p.m.! Both men set up telephone companies and became bitter rivals. Eventually, Gray sold his Western Union Telegraph Company to Bell.

Bell had great success with his invention. He demonstrated it at exhibitions and people were interested in this new method of communication. The next problem was how to link up telephone lines so that people could ring up anyone they liked. The answer was a telephone exchange, where lines from different telephone subscribers could be plugged into a switchboard to connect them to each other. The first telephone exchange was set up in 1878, in New Haven, Connecticut. It connected lines between twenty-one telephone subscribers in the New Haven area, one of whom was the writer, Mark Twain (1835–1910). By 1885, there were 140,000 subscribers and 800 telephone exchanges. The number of telephone subscribers and exchanges continued to grow in other countries across the world.

The early telephone exchanges were manual, which meant that operators sat in the exchange and plugged the lines into a switchboard by hand to connect calls. Today, callers can dial almost anywhere in the world without having to go through an operator. The calls are connected by computers.

WIRELESS LINKS

The methods of communication developed so far relied on wires to link up transmitters and receivers. In the middle of the nineteenth century, scientists began to examine the idea of transmitting sounds without wires.

The first man to introduce the idea of electromagnetic waves was the British scientist, James Clerk Maxwell (1831–79), who demonstrated that light is an electromagnetic wave and suggested the

◁ *Guglielmo Marconi with his equipment for making radio transmissions.*

△ *Radios on board ships meant that even during a long ocean voyage, the passengers and crew were not completely cut off from land or from other ships.*

idea of radio waves. In 1888, German scientist, Heinrich Hertz (1857–94) produced and detected radio waves with a simple transmitter.

MARCONI'S IDEAS

But the person who actually invented the wireless, or radio, as we call it today, was an Italian electrical engineer, Guglielmo Marconi (1874–1937). In 1894, Marconi read a newspaper report describing electromagnetic waves which traveled through space at 186,000 miles (300,000 kilometers) a second. He resolved to find out if these "wireless" waves could be used to transmit sound. At first, he worked in a makeshift workshop in his parents' villa near Bologna. He built a simple transmitter and managed to detect radio waves.

Marconi realized that he could send transmissions over longer distances with a more powerful transmitter and receiver. A Russian scientist, Aleksandr Popov (1859–1905), was also studying radio waves, although he used them to detect distant thunderstorms and not for communication. In 1895, Popov discovered that a long vertical wire on the receiver made it more sensitive to the signals, and so invented the antenna. Marconi made use of this invention by fitting antennae to his transmitter and receiver. He worked on his invention until he managed to send a signal from the house to a field one mile (two kilometers) away.

ACROSS THE AIRWAVES

Tremendously excited by his discovery, Marconi offered his invention to the

(fifty kilometers) across the English Channel, between Dover on the English coast and Wimereux near Boulogne on the French coast. In 1901, with the help of an Englishman, Sir John Fleming (1849-1945), he made the first radio link across the Atlantic between Cornwall and Newfoundland.

So far, radio messages had been sent in Morse code. The human voice was first heard on the radio in the United States just before World War I. Marconi had formed his own wireless company by this

HOW SOUND AND RADIO WAVES TRAVEL

Sounds are made by rapid vibrations, or sound waves, which the brain translates so that we can understand them. Sound waves spread out like ripples on water, becoming weaker as they get further from their source. Sound travels through the air at about 750 m/hr (1,200 km/h).

Sound waves have to have something to travel through. They can travel through solid materials like walls as well as through air, but thicker materials absorb some of the sound. There is no sound in space because there is no air for it to travel through. Astronauts have to speak to each other by radio because radio waves can travel through space.

Electromagnetic waves such as light and radio waves travel faster than sound. To transmit sound by radio waves, a microphone in a transmitter converts them into electrical signals. The signals pass to an antenna in the transmitter and spread out as radio waves. The antenna on the receiver picks up the waves and a loudspeaker turns them back into sounds.

Italian Ministry of Posts and Telegraphs. To his great disappointment, they were not interested, so he went to London to demonstrate his invention. The British government was interested in his work and asked him to give a demonstration to army and navy officers. Naval officers quickly saw the possibilities of wireless signals for shipping. In 1899, Marconi transmitted a message about thirty miles

time and he continued to develop his invention. He was most interested in using radio for communication between specific groups, such as soldiers in the field or ships at sea. But his invention was soon to become a means of communicating to a far wider audience. In 1920, the Marconi Company broadcast the first British radio program and suddenly everyone wanted a wireless in their home. Before the invention of television, radio was the main form of home entertainment and also kept people up to date with world news.

IN THE MODERN WORLD

Today, all these methods of communication have moved forward in ways which their inventors could never have imagined. The telephone can link people on opposite sides of the world in seconds, and people have telephones in their cars and mobile phones which can be charged up so that they work without

electrical signals. At the other end, the signals are converted back into words and printed out. Fax is a more modern way of sending paperwork over the phone. Documents are fed into a fax machine which turns the text and pictures into electrical signals. The signals are sent along the telephone wires and a fax receiver at the other end turns them back into exact copies of the documents sent.

Radio is used as a method of communication for shipping, aircraft, the armed forces, and emergency services

wires. A telephone which shows a picture of the caller and recipient on a small screen is now becoming available.

The telex was invented in 1916 and is still used today, although it has largely been taken over by the fax, or facsimile, machine. A telex is a type of telegraph which is sent over a telephone. The message is typed out at one end and travels along the telephone wires as

such as the police. Radio now spans distances undreamed of by Marconi. Communications satellites have made it possible to link up distant corners of the world and even to communicate with craft in space.

Today, most people have a telephone in their home, but at the beginning of the century, the newly invented telephone was a luxury few people could afford.

RECORDING SOUND

*From the first crackling records
to today's superb quality compact discs,
sound recording has brought
pleasure to millions.*

The nineteenth century was a time of great change, and inventors came up with a whole range of new ideas. One of these was the concept of storing sounds on a solid material so that they could be played over and over again.

The first machine for recording and reproducing sounds was invented in 1877, by an American, Thomas Alva Edison (1847–1931), one of the greatest inventors of the nineteenth century. He made his fortune in 1871, when he invented a paper ticker-tape machine which relayed stock exchange prices and news to banks and offices.

In 1876, Edison set up his own research laboratory at Menlo Park near New York with the money from his ticker-tape machine. The following year, he invented the phonograph for recording and playing back sound. He got the idea from the telephone, which had recently been invented. He knew that the mouthpiece

△ *The tiny personal stereo, so often used today, produces sound far better than the enormous gramophones of the turn of the century.*

and earpiece of the telephone contained disks which vibrated in response to the human voice.

Edison saw that he could connect a needle to a mouthpiece with a vibrating disk. If he spoke into the mouthpiece, the disk would vibrate and the needle would cut a groove into a solid material in the pattern of the sound vibrations. A second needle fitted to the vibrating disk in the earpiece of the instrument would then convert the sounds and play them back.

In July 1877, Edison tried out his idea. He constructed a recording machine and shouted the word "hello" into it. The sound that came back to him was an indistinct but definite "hello."

A MECHANICAL EAR

Edison now set about making a more permanent recording machine which he called a "phonograph." A needle was attached to a thin skin, or membrane. This membrane was stretched across the narrow end of a horn, rather like the eardrum at the end of the ear canal. The horn collected the sounds and channeled them to the membrane, in the same way that the ear flap collects sounds and channels them to the eardrum.

When Edison spoke into the horn, the membrane vibrated with the sound of his voice. With each vibration, the needle cut tiny marks into a sheet of tinfoil wrapped round a revolving cylinder. As the cylinder turned, the needle made a pattern of marks in a continuous spiral. It cut into the foil at different depths for different sound vibrations. This is called "hill-and-dale" recording because the groove is like a series of hills and valleys.

To play the recording back, the needle moved along the spiral groove. The indentations in the foil made the needle and membrane vibrate in different ways, and the sound recording was heard through the horn.

◁ *Thomas Alva Edison.*

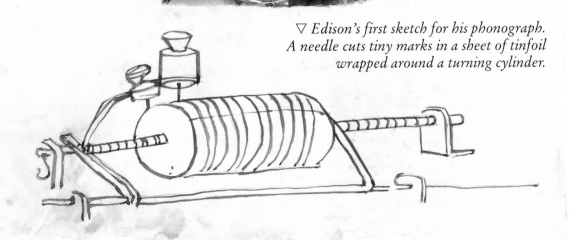

▽ *Edison's first sketch for his phonograph. A needle cuts tiny marks in a sheet of tinfoil wrapped around a turning cylinder.*

Edison's first recording was the nursery rhyme "Mary had a little lamb."

Edison's invention brought him worldwide fame, but his phonograph was very basic. The sound quality was not good and the cylinders did not last long. However, Edison was more interested in thinking up new ideas than in perfecting his inventions. He took out more than 1,000 patents for his various inventions. Among other things, he invented a light bulb and a storage battery, and, in 1912, he made the first talking motion pictures.

It was left to others to improve on the phonograph. In 1886, a machine called a "graphophone" was invented. It was similar to the phonograph but used a wax cylinder instead of tinfoil and a cutter shaped like a chisel instead of a needle. This gave better sound quality than Edison's machine.

THE FIRST DISCS

In 1887, Emile Berliner (1851–1929), a German living in America, came up with an idea which we still use today: the record. Berliner's record was a flat disk made of zinc covered with a layer of wax.

On the phonograph, the person spoke into the horn to make a recording. The horn also amplified the sounds when they were played back. This worked reasonably well for recording a single person talking or singing, but it was not so good for recording several people such as a group of musicians.

A cutter made a spiral groove in the wax, but instead of up-and-down indentations like hills and valleys, it cut a wavy groove which ran from side to side along the surface. This method, known as "lateral" recording, reproduced sounds more accurately than the hill-and-dale method. When the groove had been cut into the wax, acid was used to etch it into the metal disk. This created a permanent master copy from which other copies could be made. Flat disks could be pressed so that both sides of the record were stamped at once, which made them far easier to copy than cylinders.

The records had to be copied onto a soft material which set hard. Early records were made from shellac, a natural plastic made from the secretions of the lac insect found on some trees in India and Thailand. Today, records are made from vinyl plastic.

Having found a way to make records, Berliner went on to design a machine for playing them: the "gramophone." The early gramophone had a large horn for making the record sound louder. The machine had to be wound up with a handle to make the turntable revolve.

LOUDER AND CLEARER

The sound on a gramophone with a horn was crackly and not very realistic. The next big improvement was to introduce a microphone and loudspeaker into the gramophone.

There are different types of microphones, but they are all used to change sound into a varying electric

Emile Berliner invented the record which is still used today.

current. One of the earliest microphones was invented in 1878 for use in the telephone. This is known as the "carbon" microphone. Vibrating disks in the telephone form the lids of two small containers filled with grains of carbon. When each disk, or "diaphragm," vibrates as someone talks, the carbon moves in the container.

The speed of the vibrations depends on the pitch of the person's voice—how high or low it is. The amount of vibration depends on the volume of the voice. When the diaphragm moves inward, the grains of carbon are packed tightly together. This allows more electricity to flow through the carbon. When the diaphragm moves outward, the grains are packed more loosely and less electricity flows through. So a varying sequence of signals is passed down the wire.

THE MOVING COIL

The microphone used for recording is the "moving-coil" microphone. A thin disk of metal vibrates when sounds hit it. Attached to the metal is a coil of wire which also moves. A magnet produces an electric current in the wire. The strength of the current varies depending on the pitch and volume of the sounds.

A machine with a microphone also needs a loudspeaker to play back the sounds. A loudspeaker has a motor which usually consists of a coil of wire moving near a magnet. The electric current from the microphone is fed into the motor, and the signals cause a diaphragm to vibrate. The vibrations make the air around them

△ *Tiny variations in the groove of a record cause the stylus to vibrate in different ways. The vibrations of the stylus are then amplified to produce a copy of the original sound.*

vibrate too, and this reaches our ears as the recorded sounds. The signals from a microphone are very weak and have to be strengthened by an amplifier before they are passed to the loudspeaker. Early microphone and loudspeaker systems were not nearly as sophisticated as today's, but even the earliest versions improved the quality of sound recording.

TAPE RECORDERS

The earliest type of magnetic tape recording was patented in 1898 and stored electrical signals on magnetized steel wire. However, the idea did not catch on because there was so much electrical interference that it was hard to make out

what the sounds were meant to be. A better type of tape was introduced in the United States in 1927. This was a strip of paper covered in a liquid containing very tiny iron filings. When the liquid dried, the iron filings remained on the surface. As the tape moved past the "record head" during recording, electric currents magnetized the iron dust. The strength of the current varied depending on the sounds being picked up and so created a magnetized pattern on the tape. When the tape was played back, another magnet called the "playback head" converted this pattern into the recorded sounds.

This system was developed in Britain, the United States, and Germany during the 1930s. By 1936, it was efficient enough to use for the first tape recording of a concert orchestra in Berlin.

Early tape recorders were large machines known as "reel-to-reel" tapes. Each tape was wound on to a single large spool. To use the tape, the loose end had to be threaded past the magnetic heads and wound on to a second "take-up" spool. Reel-to-reel tape recorders are still used for some things, such as professional sound recording, but a more modern idea for home use is the cassette tape recorder, which was first introduced in 1961. The tape is mounted on two small spools inside a plastic case, the cassette. The whole cassette is pushed into the machine to record or play the tape.

HI-FI SOUND

Today, people can listen to the music of a complete orchestra, with the sound of each instrument faithfully reproduced. The accuracy of a recording is known as the "fidelity." Equipment which produces high quality and accurate sound reproduction is known as high-fidelity or "hi-fi" equipment.

Most music is now produced using "stereophonic" or "stereo" sound. Several microphones are used to record the music. Each microphone is fitted to a separate amplifier on a control desk and picks up a different aspect of the music. The recorded music is played back through two or more speakers. Stereo gives the effect that the music is all around you, as if you are in the audience at a concert. Sound has come a long way since Edison's first squeaky recording of "Mary had a little lamb."

▽ *A vast range of equipment for recording and playing back sound is available today.*

DIGITAL RECORDING

The latest method of reproducing music is digital recording, which was introduced in the 1980s. Sound is stored as a digital code which is translated into sound by a computerized player. The compact disc, or CD, is the best-known form of digital sound recording. It is a thin plastic disc covered with a shiny protective coating. The surface of the disc is covered with an arrangement of very tiny pits in a special code. When the disc is played in a CD player, a laser beam reads the code and changes it into electrical signals which are then played as sounds. Compact discs produce the finest-quality sound available at the moment. There is no crackling and manufacturers claim that they do not wear out, however much you play them.

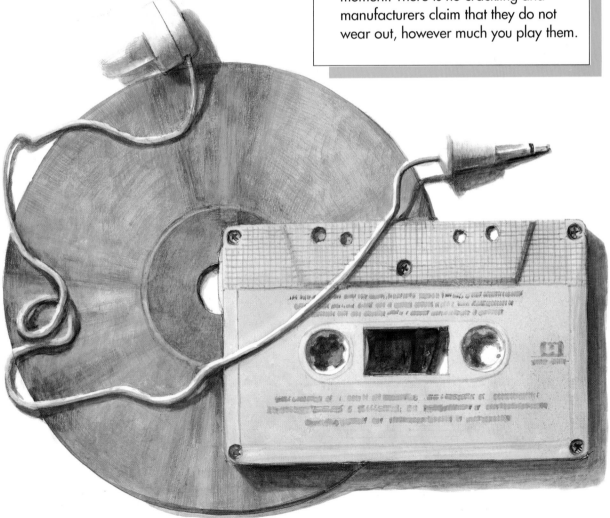

MOVING IMAGES

We think of moving pictures as a very modern invention, but the first steps towards today's films, television, and videos were taken over 150 years ago.

When you go to see a film or watch television, you are witnessing the inventions of many different people. We take these ideas for granted today, but, in the nineteenth and early twentieth century, the concept of moving pictures was very new and exciting.

The idea of moving pictures actually came before photography had been

△ *Early cameras were bulky to carry about and difficult to set up.*

◁ *William Fox Talbot.*

invented. One of film's ancestors was the magic lantern invented in 1654. This was a way of using light to show still pictures. In 1829, a French scientist, Joseph Antoine Ferdinand Plateau (1801–83) worked out the theory which forms the basis of moving images. His intention was to prove that the eye can be tricked into seeing something that is not accurate.

An action such as running is broken down into a sequence of step-by-step pictures. The pictures are then played back at the same speed as the original movement. Because the human eye retains each image for about one-thirtieth of a second, the pictures all merge in the viewer's mind and create the illusion of movement. In 1832, Plateau made a device called a "phenakistiscope" to demonstrate his theory. A revolving cardboard disk contained the sequence of pictures showing each tiny stage of a movement. Each of the pictures was revealed to the viewer in turn, giving the impression that the image was moving.

CAPTURING AN IMAGE

These early moving pictures were hand-drawn images. But the development of photography was to have a huge impact on the idea of moving pictures. To make

THE CAMERA OBSCURA

The principle on which photography is based had been known since ancient times. A beam of light coming into a darkened room through a small hole in one wall will project a reversed, upside-down image of the scene outside on the opposite wall. Archimedes (c. 287–212 B.C.), the Greek mathematician and inventor, knew of this principle, known as the "camera obscura," Greek for "dark room." The Arab scholar Alhazen (965–1038) used it to watch eclipses of the sun, and Leonardo da Vinci (1452–1519) described the principle in great detail. The camera obscura was first used to project temporary, hand-drawn images in the late sixteenth and early seventeenth centuries.

a photograph, a projected image had to be captured on a substance which could retain it permanently. People had known how to project images for centuries, so they only needed to discover how to fix the image. Scientists knew that silver nitrate turned black when it was exposed to the light, so at the beginning of the nineteenth century, several people experimented with using silver salts to fix images on paper. In 1802, the first images were produced on paper coated in silver nitrate, but the pictures quickly faded.

A Frenchman, Joseph Niepce (1765–1833), was the first person to capture a permanent image. He began to experiment in 1816 and, after many attempts, he managed to make a picture on a sheet of paper coated with silver chloride at the back of a camera obscura. But the image was a negative with all the tones reversed. He did not succeed in making a print from his negative, but instead concentrated on finding a way of making a positive image straightaway. In 1826, he finally managed to produce a picture on a metal plate.

The following year, Niepce met another Frenchman, Louis-Jacques-Mandé Daguerre (1787–1851), and the two began to work together. In 1835, Daguerre perfected a process of producing an image on a metal plate. The quality of his "daguerreotypes" was exceptional, and they became very fashionable during the 1840s and 1850s.

A PERMANENT PRINT

But it was William Henry Fox Talbot (1800–77) who made the most significant step towards modern photography. Fox

▷ *The magic lantern was the earliest method of projecting pictures on to a screen. The pictures were hand-drawn stills.*

Talbot wrestled for six years with the problem of converting a negative to a permanent print on paper. In 1839, he at last managed to obtain a positive print on paper.

As techniques improved, photography became more popular. In 1871, a young American amateur photographer, George Eastman (1854-1932), invented a machine for efficient developing of photographic plates and founded the Eastman Dry Plate Company, which later became the world-famous Kodak. In 1888, Eastman revolutionized photography with the Kodak box camera, which was light and simple to use. For the first time, anyone could take photographs without setting up complicated equipment.

Eadweard Muybridge.

his invention to photograph the flight of seagulls on the seashore in Naples, where the locals commented on the crazy man who aimed a gun at birds without ever shooting any!

Now the possibilities began to interest the American inventor Thomas Edison. He began discussions with Muybridge about combining his phonograph with the "Zoopraxiscope," a machine which Muybridge had designed for projecting his photographic sequences. In 1888, Edison's team at Menlo Park came up with their own device for viewing films, which they called the "Kinetoscope."

THE PICTURES MOVE

Meanwhile, other people had seen how photography could play a part in producing moving images. A step-by-step sequence of photographs could show movements far more accurately than drawings.

The main pioneer of action photography was Eadweard Muybridge (1830–1904). Muybridge took endless pictures of moving animals and people, and proved, with one of his step-by-step series, that a galloping horse lifts all four legs off the ground at the same time, a fact that no one had realized before.

Another photographer working on movement shots was a Frenchman, Etienne Jules Marey (1830–1904). In 1882, he invented a photographic rifle that allowed him to take a series of photographs very rapidly. Marey used

They used a camera like Marey's to make the films, but they introduced an important improvement which is still used to this day. Marey had found it difficult to move the film smoothly through the camera. Edison's team used film with holes along the edges. As the film wound on, the holes engaged with a sprocket in the camera and in the playback machine.

The kinetoscope was a wooden box containing a battery-powered motor, a lamp, a magnifying lens, and an eyepiece. Only one person could watch the film at a time, peering through the eyepiece as the film flickered past the lamp.

The quality of these early films was very poor. To improve it, Edison opened the first movie studio, an open-topped shed on wheels. It had large windows and could be moved around to follow the sun, so that the scenes being played inside were always well-lit.

People were intrigued by moving pictures and, in the 1890s, they became a popular attraction. Peep-show parlors began to appear, where people could peer into kinetoscopes and marvel at the moving images.

DRAWING THE CROWDS

The interest in films inspired other people to improve on Edison's equipment. An obvious development was to find a way of showing a film on a big screen so that a crowd of people could watch it at the same time. Edison started to work on some ideas but soon lost interest. It was left to two French brothers, Louis (1864–1948) and Auguste (1862–1954) Lumière, to develop the ancestor of the movie projector.

Edison's kinetoscope and the Lumières' equipment.

The Lumière brothers ran a factory producing photographic equipment in Lyons, France. They began to take an interest in motion pictures, and in 1894, they set to work to improve on Edison's kinetoscope. On February 13, 1895, they

◁ *Walt Disney using a Pathé camera.*

STEP-BY-STEP TO MOVING PICTURES

1826 Niepce produces the first photograph on a metal plate.

1829 Plateau puts forward his "persistence of vision" theory.

1839 Fox Talbot discovers the negative/positive process and prints a photograph on paper.

1878 Muybridge produces a sequence of photographs showing how a horse gallops.

1882 Marey invents a photographic rifle for action shots.

1884 Nipkow invents the scanning disc.

1888 Edison's team patents the kinetoscope.

1895 The Lumière brothers give the first public cinema show in Paris.

1897 Pathé invents a separate camera and projector.

1923 Zworykin demonstrates the conversion of an image into electrical signals.

1926 Baird gives the first public demonstration of television.

1929 The BBC begins to broadcast television programs in Britain.

1937 Baird's system is dropped by the BBC in favor of Marconi-EMI's electronic equipment.

patented their invention, which they called the "cinematograph." For the rest of that year, they gave private film shows to arouse interest in their work. On December 28, 1895, they gave their first film show to a paying audience, in a makeshift cinema in the basement of a Paris cafe.

These early films were not very exciting. The Lumières' first film showed workers leaving their factory in Lyons, and another showed a baby eating breakfast. On the other hand, some of their films were more exciting. One film showed a train coming into a station and was shot at such an angle that the audience were terrified that the train was actually going to hit them.

But the main thing was that the pictures moved. The novelty was so great that audiences flocked to see anything.

By 1897, the Lumières had made 358 short films. By 1901, there were 1,299.

Other people soon followed the Lumière brothers and began to make film equipment. A Frenchman, Charles Pathé (1867–1957), invented a separate camera and projector in 1897, and then set up a company to make films. The film craze spread from France to America, where a huge film industry quickly began to grow. Many people contributed to the growth of this industry, including the famous Walt Disney (1901–66), who is perhaps best known for cartoons such as Mickey Mouse, first designed in 1928.

THE INVENTION OF TELEVISION

The discovery that sounds could travel by radio waves set several scientists thinking. If sounds could be transmitted in this way, what about pictures?

The main problem was how to convert a picture into a continuous sequence of information, a process called scanning. When you look at a picture on a television screen, you are seeing millions of tiny dots of light. These dots are arranged in lines, like the lines in a book. On a modern television, the picture has 625 lines. The more lines the picture has, the more detailed and precise it can be, because the image is broken up into very tiny parts.

Before a television program can be transmitted, these spots of light have to be converted into electrical signals which can be sent out on electromagnetic radio waves. This conversion is made in the television camera. The television set is a receiver. The antenna picks up the signals. The cathode ray tube in the television converts the signals into pictures, and the speaker produces the sounds.

Some of the first pioneers of television looked at electronic methods. In 1923, a

△ The famous comic actor Charlie Chaplin (1889-1977) made his first film in 1913 and was an immediate success with his character, the tramp in baggy trousers and bowler hat. Chaplin became one of the great stars of the early silent films.

Russian-born American, Vladimir Zworykin (1889–1982) developed a procedure for converting an image into electrical signals.

SCANNING THE IMAGE

Meanwhile, another pioneer was developing television in quite a different way. John Logie Baird (1888–1946) was a Scottish engineer turned inventor. He had been struggling with unsuccessful inventions for nine years when, one day in 1923, he came up with an idea for a mechanical scanning system. It was based on an 1884 invention by a German student, Paul Nipkow (1860–1940). Nipkow had had the idea of cutting up images into lines. He produced an "electric telescope" which consisted of a disk pierced with a spiral of holes. When the disk spun in front of an object, it divided the object into a series of lines.

A FLICKERING IMAGE

Baird applied for a patent to use Nipkow's disk to make a television with a mechanical system of scanning. He used old radio parts and other bits and pieces to make a clumsy transmitter, and, in 1925, he managed to produce a flickering, shadowy picture.

Greatly excited by his success, Baird borrowed money to improve his equipment and rented an attic workshop in London. In 1926, he televised a ventriloquist's dummy at a public demonstration in his workshop. The image was made up of only eight lines so it was very dim and blurred. But, despite the poor quality of the picture, the demonstration was a great success. Like the first films, the idea of television was so new that any success seemed like a miracle. People became very excited by the possibilities of this new communication system.

Baird set up his own company to work on improving the quality of his pictures. The BBC had been founded in 1922 to broadcast radio programs, and in 1929 Baird persuaded them to transmit a television service.

Meanwhile, other companies were working on the electronic methods begun by Zworykin and others. In 1933,

John Logie Baird's clumsy contraption used a spinning disk with holes round the edge to break an image up into lines.

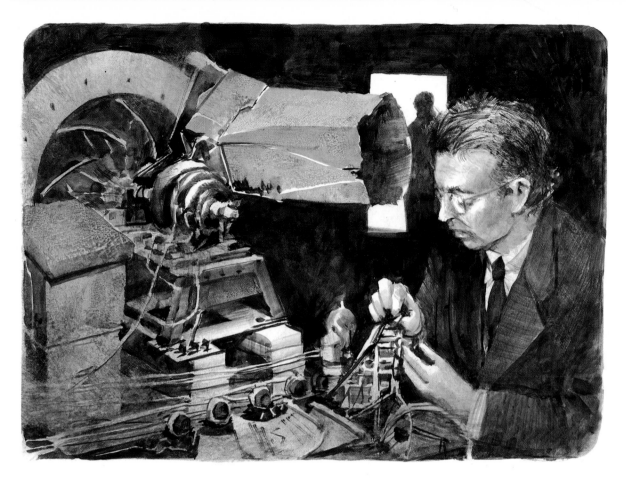

Zworykin patented the "iconoscope," a camera tube that converted an image into electrical pulses. It was used by RCA (Radio Corporation of America) for experimental broadcasts in 1936. In Britain, the Marconi Company and EMI (Electrical Musical Instruments) were working on improved television systems.

BETTER QUALITY PICTURES

The equipment they came up with was more sophisticated than Baird's mechanical device, which had several drawbacks. The camera was fixed in position so anyone being televised had to stay in one position. The quality of the picture was so bad that people being televised had to wear clown-like makeup so that their features showed up, and the flickering light from the machine almost blinded them. The last straw was that Baird's system kept on breaking down.

△ *Despite the very poor quality of the picture he produced, Baird's invention caused great excitement when he demonstrated it at his workshop in Soho, London, in 1926.*

The BBC now had enough choice to set higher standards for their programs. Baird invented a 240-line mechanically scanned system which was an improvement, but Marconi-EMI introduced a 405-line system with electronic scanning. For a while, the BBC used both systems for their programs, but in 1937, they finally dropped Baird's system in favor of the electronic method.

Baird was bitterly disappointed by this rejection. But, even though other pioneers were working on ideas and introducing more sophisticated equipment than Baird's, those first flickering pictures he produced in 1925 earned him credit as the inventor of television.

THINKING MACHINES

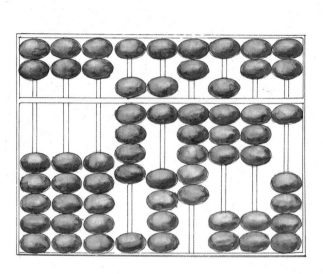

The invention that started as a way to make mathematics simpler has now developed into a vast network of machines used in almost every aspect of modern life.

For almost as long as people have been doing mathematical calculations, they have been looking for ways of making them easier to carry out. The task of adding up long lists of figures was no more appealing in ancient times than it is today, and there was always a danger of getting it wrong.

The earliest counting device was the abacus, invented in about 3000 B.C., probably in Babylonia. In the simplest form of abacus, pebbles representing thousands, hundreds, tens, and units were moved about in sand. The Romans had a version in which pebbles or beads were moved along slots in a metal plate. The modern abacus, in which beads are slid along wires in a frame, is still used in China, Japan, India, and Russia.

The abacus was the ancestor of the calculator, but it only really helped by grouping pebbles, counters or beads so

△ *The abacus, first invented 5,000 years ago but still used today.*

▷ *Early calculators: a version of the abacus (top); Napier's bones (center); Blaise Pascal's mechanical calculator (bottom).*

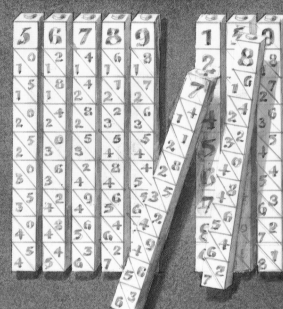

Centaines de Mille	Dixaines de Mille	Miller	Centaines	Dixaines	Nombres Simples	Sols

that they could be counted more easily. It did not do the work for you, as modern calculating machines do.

CALCULATING MACHINES

It was several centuries before anyone attempted to invent a machine which did this. In 1624, Wilhelm Schickard (1592–1635), a German professor at the University of Heidelberg, invented a machine which could add, subtract, multiply, and divide. He called his invention a "calculator clock."

But the first true calculating machine was made in 1642 by the French philosopher Blaise Pascal (1623–62).

Pascal was only nineteen when he invented a machine to help his father, who was a tax collector. The calculator, which he named the "Pascaline," used a system of gears to add or subtract figures. It could tackle up to eight columns of figures at a time. Later in the seventeenth century, the German mathematician Gottfried Leibniz (1646–1716) invented a machine similar to Pascal's, which could also multiply and divide. Both machines

▽ *Babbage's "Analytical Engine" worked on a system of punched cards which were adapted from cards used to control colored threads on a loom.*

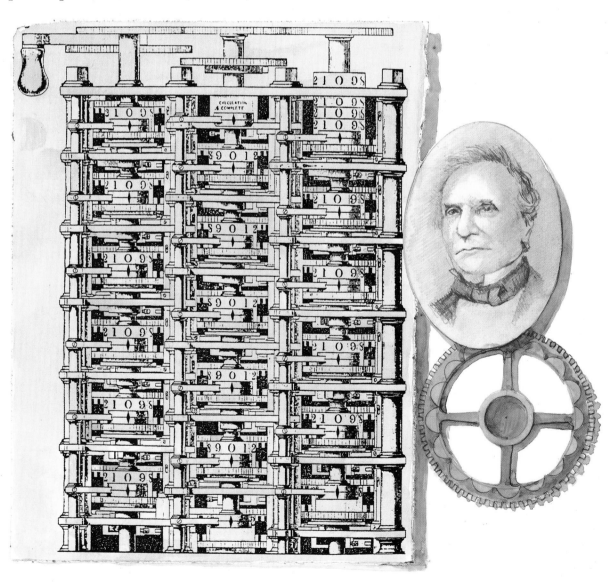

used the same technique, known as "single step" calculation. This repeats the same operation, such as a series of additions, to arrive at the correct answer.

People still needed devices to help them carry out complicated calculations such as long multiplication and long division. One mathematical aid was "Napier's Bones," invented by the Scottish mathematician John Napier (1550–1617). This consisted of a set of metal rods with numbers on them. When the rods were lined up together in a certain way, they formed a multiplication table. Long multiplication calculations could be carried out by following the table and making some simple additions.

Napier also invented "logarithms" in 1617. This was a book of tables which converted multiplication to a series of additions, and division to a series of subtractions. By looking up the logarithms of numbers, and adding or subtracting them, even the most complicated calculations can be carried out quite simply.

THE ANCESTOR OF THE COMPUTER

These early devices were helpful but they did not begin to fulfill the work of the modern computer. The first machine to do this was invented by an Englishman, Charles Babbage (1792–1871). Babbage was professor of mathematics at Cambridge University when he began to experiment with computing machines. His "Analytical Engine," which he began to develop in 1835, carried out calculations and stored data and results. The data was fed into the machine on punched cards. These instruction cards, which were really early versions of computer programs, were written by Babbage's colleague, Ada, countess of Lovelace (1815–52).

△ Ada, countess of Lovelace, has been called the first computer programmer.

The designs Babbage produced show similar features to the computer, although it was a mechanical rather than an electronic device. Lady Lovelace also realized that the machine would often have to carry out the same functions and discovered a way of producing repeat instructions for the machine to follow, so that it was not necessary to punch cards with instructions that had been used before. The same principle is used in computer programming today.

Babbage never built a finished version of his machine, because he was always more interested in moving on to the next project than in completing a current one. Because he did not follow it through, few people knew about his machine, and it was many years before anyone began to delve into these theories again.

PIONEERS OF THE MODERN COMPUTER

In the mid-1930s, groups of people in Britain and the United States began to investigate the idea of developing a computer. In Britain, the mathematician Alan Turing (1912–54) published a paper describing a type of computer and the sort of problems it could work out. This paper was only theory, but Turing had the opportunity to put his ideas into practice when he joined the British Intelligence Service during World War II. One of his tasks was to build a computer that would crack enemy codes.

Meanwhile, in the United States, Howard Aiken (1900–73) of Harvard University began work on an enormous calculator which he built with the help of the International Business Machines Corporation.

The work of these two men and their colleagues led to the introduction of the world's first true computers, the British Colossus and the American Automatic Sequence Controlled Calculator (ASCC). Both these machines were designed to carry out mathematical calculations much more quickly than a person could do them. In fact, the rate of the calculations seems very slow by today's standards. The first ASCC could add two numbers together in 0.3 seconds, but this rate soon improved. By 1947, the Mark II model could perform the same task in only 200 milliseconds.

These early computers had many mechanical parts. The first electronic computer appeared in the United States in 1946. It was called the Electronic Numeric

△ *Alan Turing.*

Integrator and Calculator, or ENIAC. The ENIAC could perform tasks far more quickly than the earlier machines, although it still used punched cards for data. This was a laborious process compared with today's method of using a keyboard and disks. Even so, in one day ENIAC was able to complete calculations that would take a human being a whole year to do.

▷ *The Fairchild computer circuit of 1961 was the first to be manufactured commercially. On the right is a short piece of punched paper tape which was used to enter data before the days of disks. Information had to be typed onto cards or paper tape where it appeared as a series of holes. The punched card or paper was then loaded into the reading part of the computer.*

PREPARING COMPUTER PROGRAMS

A computer needs a memory or storage area where it keeps the data it is working on. It also needs to be told exactly how to do each task. Unlike the human brain, a computer cannot think for itself, so it has to be given very detailed instructions about everything it is asked to do. The set of instructions telling it how to do a particular task is called the "program." Programs stored on disks are known as computer "software." The computer equipment is called the "hardware."

Data is typed into the computer using a keyboard. The data appears on a screen or "visual display unit" (VDU) linked to the keyboard, so the operator can keep a constant check for mistakes.

Information stored on disks can also be shown on the screen.

BINARY LOGIC

A computer processes information in the form of groups of binary numbers. In the binary system, there are only two digits, 0 and 1, and each number is made up of combinations of these two symbols. So 2 is written as 10, 3 as 11, 4 as 100, 5 as 101, and so on. This is ideal for computers because each symbol can be represented by the on or off position of electronic switches. An electrical pulse indicates 1, and no pulse indicates 0.

Binary logic was actually developed by an English logician, George Boole (1815–64), as early as 1859. The ENIAC used binary numbers, but a series of lectures given by the American

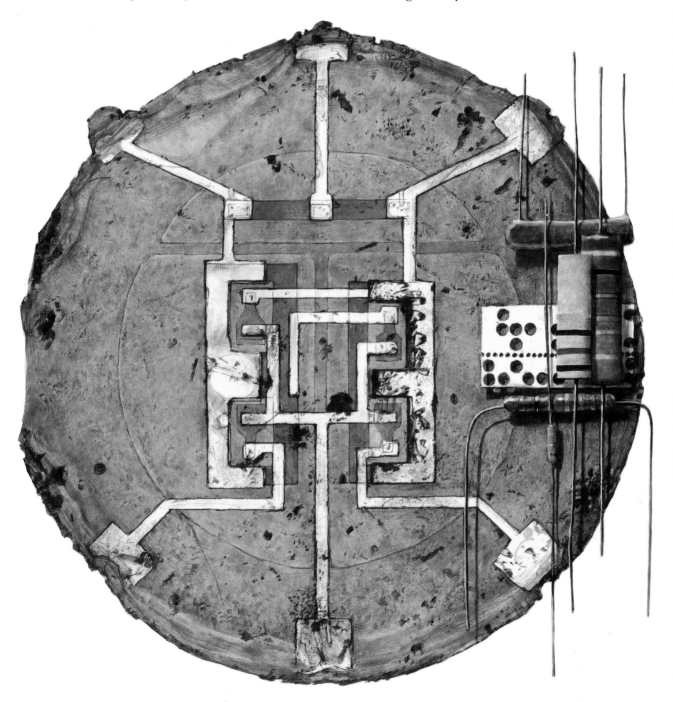

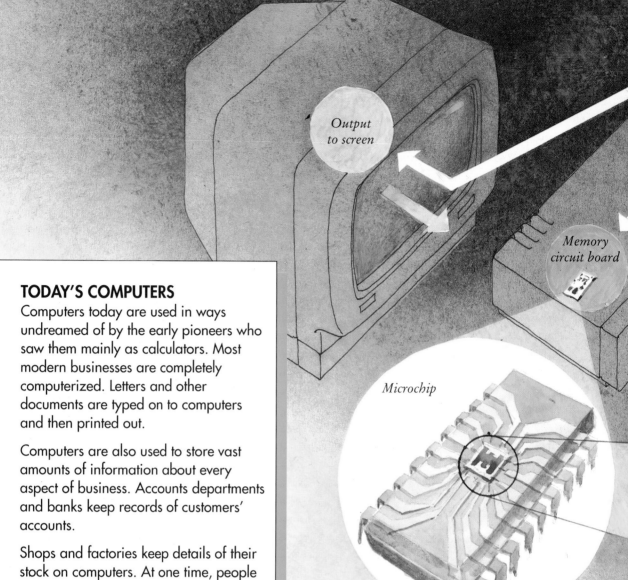

Output
to screen

Memory
circuit board

Microchip

TODAY'S COMPUTERS

Computers today are used in ways undreamed of by the early pioneers who saw them mainly as calculators. Most modern businesses are completely computerized. Letters and other documents are typed on to computers and then printed out.

Computers are also used to store vast amounts of information about every aspect of business. Accounts departments and banks keep records of customers' accounts.

Shops and factories keep details of their stock on computers. At one time, people had to check the stock so that they knew when anything was running out and needed to be reordered. Computers in shop cash registers can now keep a record of every item that is bought and send out an order automatically when stocks are running low.

Microchips are also used to give instructions to machines in less obvious ways, where there is no actual computer to be seen. For example, the robots which carry out work in factories are controlled by microchips. So are household appliances which can be programmed, such as washing machines, dishwashers and central heating systems.

mathematician, John von Neumann (1903–57), in 1946 did a great deal to emphasize the importance of this system.

ON AND OFF

The binary system needed a large number of switching devices to turn the current on and off. In the early days, this was done by thousands of valves inside the computer. The valves used a great deal of electricity, and the machine got very hot. Computers also had to be very large and cumbersome to accommodate these valves. The next development was the transistor which was introduced in 1948

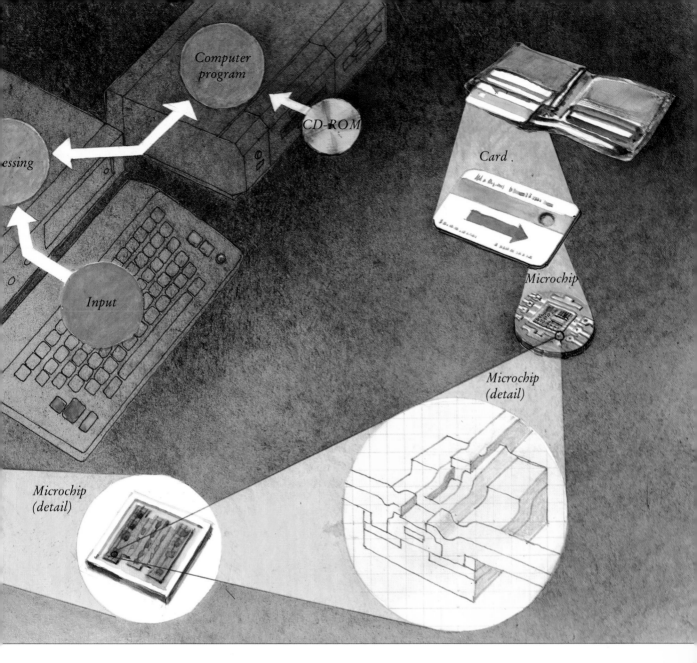

Computer
program

CD-ROM

essing

Input

Card

Microchip

Microchip
(detail)

Microchip
(detail)

by William Shockley (1910–89) and other scientists at the Bell Telephone Company. Transistors are lighter, stronger and more robust than valves. During the 1950s, they began to replace valves in all electronic equipment. The introduction of the transistor radio was one important development. Another was a smaller and more reliable computer.

SILICON CHIPS

Computers became even smaller in about 1964, when manufacturers discovered how to squeeze the components for different functions on to minute pieces,

△ *Microchip technology makes it possible for personal details on a credit card or a security pass to be checked against a master file on a central computer.*

or "chips," of silicon. In the last twenty years or so, computers have become smaller, cheaper, faster and more efficient.

At one time, only governments or large businesses owned computers, but today many people have computers in their homes. The move to the computerized world we now live in has perhaps been one of the most dramatic developments in the history of inventions.

FURTHER READING

Asimov, Isaac. *How Did We Find Out about Computers?* New York: Walker & Co., 1984.

Bostrom, Ronald. *Cameras*. Milwaukee, WI: Raintree, 1985.

Brookfield, Karen. *Writing*. New York: Dorling Kindersley, 1993.

Caselli, Giovanni. *The First Civilizations*. New York: Peter Bedrick Books, 1983.

Chisholm, Jane and Anne Millard. *Early Civilization*. London: Usborne, 1991.

Cousins, Margaret. *The Story of Thomas Alva Edison*. New York: Random House, 1981.

Endacott, Geoff. *Discovery & Inventions*. New York: Viking, 1991.

Farr, Naunerle C. *Thomas Edison - Alexander Graham Bell*. West Haven, CT: Pendulum Press, 1979.

Fodor, Ronald V. *Gold, Copper, Iron: How Metals Are Formed, Found, & Used*. Hillside NJ: Enslow, 1988.

Gleasner, Diana. *The Movies*. New York: Walker & Co., 1983.

Gonen, Rivka. *Pottery in Ancient Times*. Minneapolis, MN: Lerner, 1974.

Haney, Jan P. *Calculators*. Milwaukee, WI: Raintree, 1985.

Lampton, Christopher. *Astronomy: From Copernicus to the Space Telescope*. New York: Franklin Watts, 1987.

Lebrun, Francoise. *The Days of the Cave People*. Lexington, MA: Silver, Burdett & Ginn, 1986.

Macaulay, David. *The Way Things Work*. New York: Dorling Kindersley, 1988.

Platt, Richard. *Cinema*. New York: Dorling Kindersley, 1992.

Reid, Struan. *Usborne Illustrated Handbook of Invention and Discovery*. London: Usborne, 1986.

Smith, Elizabeth. *Paper*. New York: Walker & Co., 1984.

Wood, Tim. *Prehistoric People*. New York: Franklin Watts, 1980.

INDEX

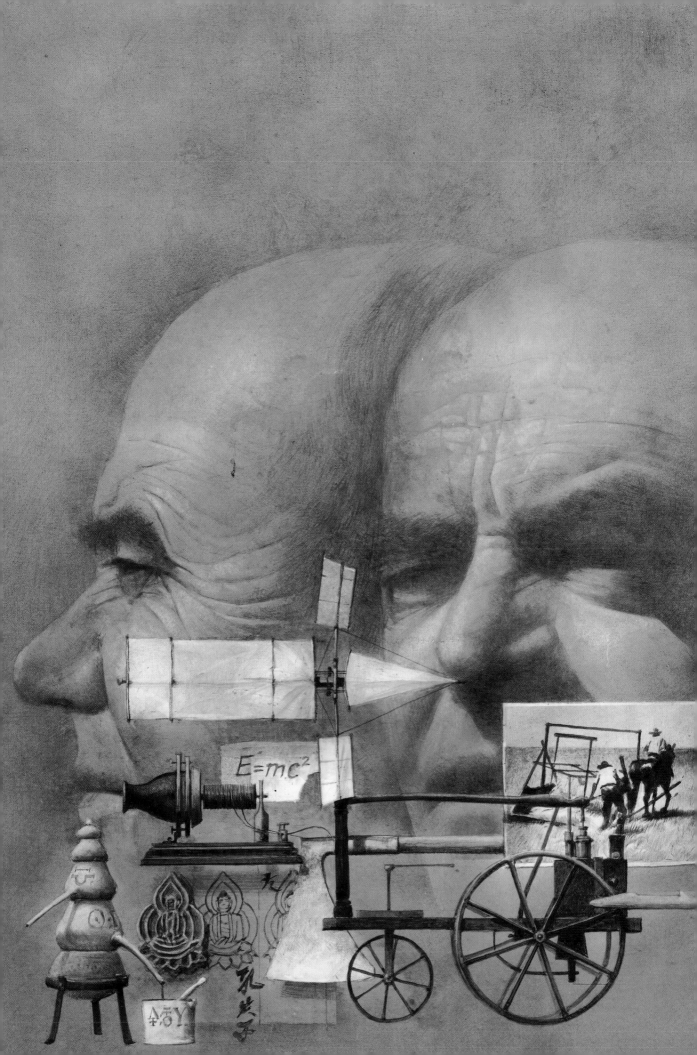